Cucurbiturils

Chemistry, Supramolecular Chemistry and Applications

Other Related Titles from World Scientific

Cryptands and Cryptates
by Luigi Fabbrizzi
ISBN: 978-1-78634-369-7

Hydrogen-bonded Capsules: Molecular Behavior in Small Spaces
by Julius Rebek, Jr.
ISBN: 978-981-4678-35-3

Cyclodextrin Chemistry: Preparation and Application
edited by Zheng-Yu Jin
ISBN: 978-981-4436-79-3

*Handbook of Porphyrin Science: With Applications to Chemistry,
Physics, Materials Science, Engineering, Biology and Medicine —
Volume 1: Supramolecular Chemistry*
edited by Karl M. Kadish, Kevin M. Smith and Roger Guilard
ISBN: 978-981-4280-17-4

Cucurbiturils

Chemistry, Supramolecular Chemistry and Applications

Kimoon Kim

Institute for Basic Science, South Korea & Pohang University of Science and Technology (POSTECH), South Korea

James Murray

Institute for Basic Science, South Korea

Narayanan Selvapalam

Kalasalingam University, India & Pohang University of Science and Technology (POSTECH), South Korea

Young Ho Ko

Institute for Basic Science, South Korea & Pohang University of Science and Technology (POSTECH), South Korea

Ilha Hwang

Institute for Basic Science, South Korea & Pohang University of Science and Technology (POSTECH), South Korea

World Scientific

NEW JERSEY · LONDON · SINGAPORE · BEIJING · SHANGHAI · HONG KONG · TAIPEI · CHENNAI · TOKYO

Published by

World Scientific Publishing Europe Ltd.

57 Shelton Street, Covent Garden, London WC2H 9HE

Head office: 5 Toh Tuck Link, Singapore 596224

USA office: 27 Warren Street, Suite 401-402, Hackensack, NJ 07601

Library of Congress Cataloging-in-Publication Data

Names: Kim, Kimoon, author.

Title: Cucurbiturils : chemistry, supramolecular chemistry and applications /
   by Kimoon Kim (Institute for Basic Science, South Korea & Pohang University of Science and
   Technology (POSTECH), South Korea) [and four others].

Description: New Jersey : World Scientific, 2018. | Includes bibliographical references and index.

Identifiers: LCCN 2018005494 | ISBN 9781848164086 (hc : alk. paper)

Subjects: LCSH: Macromolecules. | Supramolecular chemistry.

Classification: LCC QD381 .C83 2018 | DDC 547/.7--dc23

LC record available at https://lccn.loc.gov/2018005494

British Library Cataloguing-in-Publication Data

A catalogue record for this book is available from the British Library.

For any available supplementary material, please visit
http://www.worldscientific.com/worldscibooks/10.1142/P648#t=suppl

Desk Editors: Anthony Alexander/Jennifer Brough/Koe Shi Ying

Typeset by Stallion Press
Email: enquiries@stallionpress.com

To our families for their continued support over the years...

Preface

This book is an introduction to cucurbiturils (CBs), in which we cover their preparations, properties and applications. In many ways, CB chemistry is still an emerging research field; nevertheless, we now have a reasonably good understanding of what CBs can do, and what they can do better than other host molecules. More importantly, CB–guest pairs have begun to assert themselves in the biological sciences where their signature high-affinity binding has allowed them to compete with biological binding pairs.

In terms of research output, there are now, on average, three publications a week involving cucurbiturils. The number of research groups utilising this unique family has also grown such that a dedicated CB meeting can be supported every two years. In 2007, the first CB-dedicated workshop was held in Maryland, USA, and a dedicated CB conference has been held every two years at various locations ever since, each larger than the last. Reflecting this growth, there are a growing number of reviews of the CB field from various perspectives, but hitherto no CB book has been published. Therefore, we feel that this book is a timely overview of the CB field that we hope will inspire the readers to learn about CBs and generate research ideas that utilize them. Similarly, we hope that this book will introduce CBs to established researchers from other fields and help them envision novel uses of CBs in their own areas.

The first part of the book covers the foundations of CB chemistry including synthetic aspects of CB chemistry, their structures and properties. In particular, we describe the remarkable characteristics of the cucurbiturils family and the host–guest chemistry of each family member and compare and contrast those with other prominent host molecules. In the second part of this book, we describe specific applications of cucurbiturils in the fields of chemistry, supramolecular chemistry, materials science and biology, with a special emphasis on high-affinity guest molecules. We complete the book with our thoughts on what the outstanding challenges in the field are. It is our hope that this chapter will inspire new researchers and researchers from other fields to bring new ideas and solutions to these challenges.

We wrote this book with first-year graduate students and advanced undergraduates in mind. We assume knowledge of the basic concepts of supramolecular chemistry (such as non-covalent interactions and binding constants) and we focus on describing CB as a prominent host molecule and highlighting their remarkable characteristics, not only in the context of other host molecules, but also in comparison with natural receptors.

In the process of writing this book, we were impressed with the range of fields that CBs have been used in; many of these could not have been envisioned back when the CB family was first expanded back in the year 2000. They have been utilized to great effect as catalysts, sensors, components in self-assembly, porous materials, photonic materials, polymeric materials, soft materials and drug delivery vehicles, amongst others. In writing this book, we have learned a lot about the directions that the field is heading in, so we feel sure that this book will be useful to experienced CB researchers too, as well as to newcomers to the field.

While we have tried to be fair, in the interest of space we also had to be judicious in what we have included since we did not set out to write a comprehensive book. We simply aim to introduce CB chemistry to the reader and survey the fields that they have been employed in. We tried our best to catch all of the important papers, but may have missed some published in the latter half of 2017 while the final draft was in preparation.

KK would like to thank all of his group members, past and present, for their contribution to cucurbituril research, especially Jaheon Kim, Sang Yong Jon, Dongmok Whang, Jae Wook Lee, Hee-Joon Kim, Woo Sung Jeon, Kyeng Min Park and Dongwoo Kim. He would also like to share his gratitude to the following collaborators without which the group could not have accomplished nearly as much as it has: Kentaro Yamaguchi,

Eiichi Nakamura, Kristin Bartik, Jong-sang Park, Parimal K. Bharadwaj, Chongmok Lee, Taisun Kim, James C. Fettinger, Lyle Isaacs, Yoshihisa Inoue, Mikhail V. Rekharsky, Angel E. Kaifer, Michael K. Gilson, Cornelia Bohne, Nobuhiko Yui, David H. Thompson, Gon Seo, Wokyung Sung, Chan Gyung Park, Sung Ho Ryu, Sei Kwang Hahn, Hugh I. Kim, David V. Dearden, Nam Ki Lee, Joon Hak Oh, Young Min Rhee, Won Jong Kim, In Su Lee and Joon Won Park. The support and advice from Sir J. Fraser Stoddart, Jean-Pierre Sauvage, Jean-Marie Lehn, Roeland Nolte, Seiji Shinkai and Su-Moon Park are greatly appreciated. Financial support from Korea National Research Foundation and the Institute for Basic Science is gratefully acknowledged.

Finally, we would like to thank Jaehyoung Koo for preparing some of the figures in this book, and Kangkyun Baek, Kyeng Min Park, Annadka Shrinidhi, Tumpa Gorai, and Rahul Dev Mukhopadhyay for proofreading several chapters and providing valuable feedback.

Kimoon Kim
James Murray
N. Selvapalam
Young Ho Ko
Ilha Hwang

About the Authors

Kimoon Kim received his B.S. degree from Seoul National University (1976), M.S. degree from Korea Advanced Institute of Science and Technology, and Ph.D. from Stanford University (1986). After two years of postdoctoral work at Northwestern University, he began his academic career at the Department of Chemistry, Pohang University of Science and Technology (POSTECH) in 1988 where he now is a Distinguished University Professor. Since 2012, he has also been the Director of the Center for self-assembly and Complexity (CSC), Institute for Basic Science (IBS). His research focuses on developing novel functional materials and devices based on supramolecular chemistry. In particular, he has pioneered research on cucurbiturils and published 130 papers on these molecules since the mid-1990s. His work has been recognized by a number of awards including the Korea Science Prize (2001), TWAS award in Chemistry (TWA, 2001), Hoam Prize (Samsung, 2006), Best Scientist and Engineer Award (Korean Government, 2008), and the Izatt–Christensen Award (2012).

James Murray studied medicinal chemistry at the University of Leeds (MChem, 2010) and received a Ph.D. from the same institution, working on self-assembled monolayers (2014). After one year as a research associate at Newcastle University (UK), he joined the Center for Self-assembly and Complexity in 2016, where his research interests are the applications of supramolecular chemistry to chemical biology.

N. Selvapalam received his M.S. (1990) and Ph.D. (1997) degrees from Mysore University and IIT Delhi, respectively. During 1997–1999, he worked at the R&D center of Cancer Institute (WIA), Chennai. After working at KAIST as a postdoctoral fellow for a year, he joined CSS in 2001. From 2001, he was a research Assistant Professor at POSTECH. In 2014, he joined Kalasalingam University where he is now an Associate Professor.

Young Ho Ko received his Ph.D. from POSTECH in 1998. After three years of postdoctoral work at the Center for Biofunctional Molecules, he moved to the Center for Smart Supramolecules in 2001. From 2009, he worked as a collegiate Associate Professor at the Division of Advanced Materials Science, POSTECH. Then, he joined the Center for Self-assembly and Complexity, IBS, in 2013, where he is now a research coordinator. His research interests encompass NMR spectroscopy of supramolecular systems and host–guest chemistry.

Ilha Hwang received his B.S. degree from Korea University (2001) and Ph.D. degree from POSTECH (2007). After one year of postdoctoral work at Stanford University, he joined the Center for Smart Supramolecules at POSTECH. In 2012, he moved to the Center for Self-assembly and Complexity at IBS, where his research interests are molecular recognition and its applications in sensors.

Contents

List of Figures and Schemes

List of Tables

Chapter 1

Introduction: History and Development

Supramolecular chemistry is a relatively new area, but an important one that has been recognised by the Nobel Prize on two occasions to: Cram, Lehn and Pedersen in 1987[1,2]; and Sauvage, Stoddart and Feringa in 2016.[3–5] Supramolecular chemistry is concerned with non-covalent interactions such as hydrogen bonding, π–π stacking, hydrophobic, electrostatic and van der Waals interactions that control molecular recognition and self-assembly. Controlling and understanding such non-covalent interactions is the central pillar of supramolecular chemistry. One of the motifs of supramolecular chemistry is the formation of host–guest complexes through molecular recognition between a molecular receptor (the host) and a ligand (guest). Some of the well-known host molecules include crown ethers, cyclodextrins and calixarenes; the structures and properties of each of these determine the guest molecules that they encapsulate.

Cucurbiturils are another family of macrocyclic host molecules that have risen to prominence more recently. They are easily accessible macrocycles that possess remarkable recognition properties. The synthesis of the first cucurbituril was reported by the German chemist, Behrend, in 1905. While no structure was determined at the time, we now know that it was Cucurbit[6]uril.[6] The Behrend report described the acidic condensation of glycoluril with formaldehyde to yield an insoluble polymeric substance, also known as Behrend's polymer. Behrend suggested that this

1

material contained "at least three molecules of glycoluril" condensed with twice as many formaldehyde molecules; based on elemental analysis, he proposed (we now know incorrectly) the formula $C_{18}H_{18}N_{12}O_6$.[6,7] He noted that this molecule was exceptionally stable, even in the presence of aggressive reagents such as $KMnO_4$. Behrend also investigated some of the properties of this molecule and found that it could form co-crystals with $KMnO_4$, $AgNO_3$, methylene blue and Congo red, amongst others.

Intrigued by the properties of this material, Mock and Freeman revisited the original synthesis in 1981, almost 80 years later, armed with modern structural characterisation methods such as X-ray crystallography and NMR.[8] They found that the structure is a highly symmetric, macropolycyclic molecule, with a hydrophilic rim on each side comprised of ureido carbonyls and a hydrophobic interior. Mock dubbed this molecule "cucurbituril" because of its resemblance to a pumpkin (Figure 1.1), since pumpkins belong to the Cucurbitaceae family. Furthermore, Mock also explored the host–guest chemistry of CB[6] by systematically examining various substrates to determine what features are important for binding. In these studies, Mock noted that alkylammonium species appeared to form a 1:1 inclusion complex with CB[6]. Subsequently, Mock demonstrated some early applications of CB[6] such as the rate acceleration of azide–alkyne click reaction within the CB[6] cavity[9,10] and an early example of a molecular switch.[11] In the 1990s, the group of Buschmann made significant contributions to the chemistry of CB[6] by exploring its dye

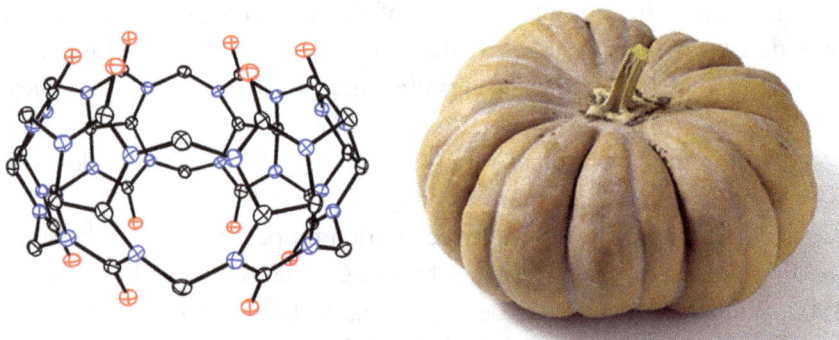

Figure 1.1.   The structure of cucurbit[6]uril compared to that of a pumpkin.

encapsulation behaviour,[12,13] its binding to cations,[14] and the calorimetric determination of binding constants for amino-functionalised molecules.[15] Kim *et al.* solubilised CB[6] in neutral media,[16] thereby expanding its host–guest chemistry and also built supramolecular assemblies using CB[6],[17] including "molecular necklace"[18] type compounds and coordination polymers. All these works will be discussed in detail later.

These early works provided a glimpse at the potential of cucurbit[*n*] urils; however, they were still relatively obscure molecules. Cyclodextrins (CD) were still the state-of-the-art molecular containers in part because CDs come in a range of sizes (six–eight sugar moieties), whereas CB[6] was the only homologue known pre-2000, so CDs had a greater range of host–guest chemistry. Cyclodextrins are water soluble too, which makes them more desirable to work with and useful for practical applications, in contrast to CB[6] which is practically insoluble in water. These are the challenges that cucurbiturils needed to meet if they were to compete with and supplant the CDs.

Development in three areas have fuelled the advancement of cucurbiturils chemistry. The first is the synthesis and isolation of other family members (*n* = 5, 7 and 8) by Kim *et al.*[19] Although CB[6] is the major product of the acid-catalysed formaldehyde glycoluril reaction, careful control of the reaction temperature allows access to different sized cucurbiturils (CB[*n*], (*n* = 5, 7 and 8)). Independently, Day *et al.* synthesised and isolated these CB homologues along with CB[5]@CB[10].[20] A note on nomenclature: Cucurbit[*n*]uril is often abbreviated as CB[*n*] or Q[*n*] or CB*n*, where *n* is the number of glycoluril units in the macrocycle. The differing volumes of the expanded family members enriches and diversifies the host–guest chemistry of cucurbiturils. CB[7] is appreciably more soluble in water than CB[6], which has helped move CB[*n*] chemistry towards biological applications. Other guests of CB[7] include redox-sensitive molecules such as ferrocene[21] and methyl viologen,[22,23] which allows external control of the complexes stability[24]; and dye molecules which enables the development of sensors.[25] The large cavity of CB[8] facilitates the formation of ternary complexes[26] which has been transformative in the field because it allows the formation of supramolecular architectures such as supramolecular polymers,[27] block copolymers[28] and nanostructures.[29] CB[8] can also serve as a reaction container for various reactions. The importance of the synthesis of the homologues can be seen in publication statistics (Figure 1.2). Since the other family members were isolated, publications have steadily increased; now there are, on average, almost three publications per week involving cucurbiturils. The number of research groups using cucurbiturils has also grown to the extent that a dedicated CB meeting or conference has been held every two years since 2007.

The second area is the high-affinity of these complexes which makes them incredibly robust such that they stand apart from other synthetic receptors. Over the years, increasingly high affinity guests (for CB[7], in particular)[30–32] have been synthesised culminating in a 10^{17} M^{-1} affinity binder for CB[7].[33] The complexes are strong enough to persist in biological media, where there are many interfering molecules. This has allowed supramolecular chemistry to be applied in biological settings, for example supramolecular hydrogels held together by CB[8][34] or CB[6],[35] and these materials remain intact in *in vivo* experiments. Guest exchange has also been demonstrated inside live cells and this has been used to activate cytotoxic nanoparticles,[36] among other applications.

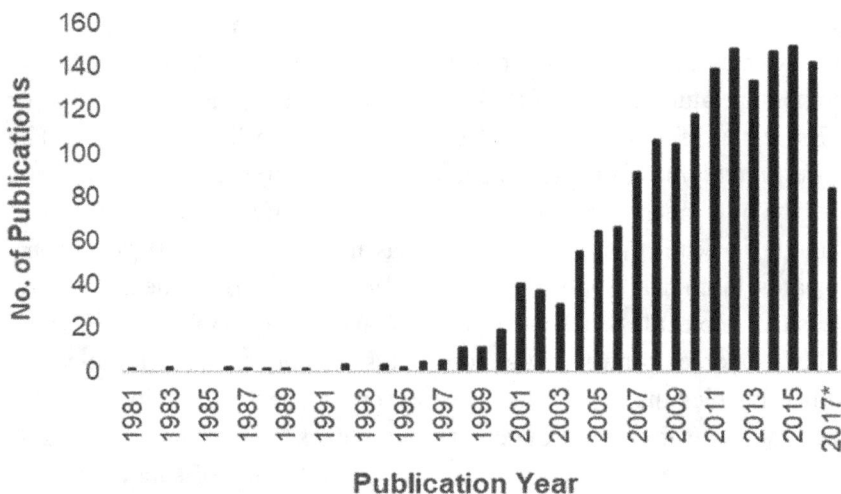

Figure 1.2.   Publications in the literature involving cucurbiturils since 1981. *Until July 2017.

The third area is the functionalisation of CBs which has allowed the remarkable properties to be harnessed in materials applications. As-synthesised CB[*n*] are quite unreactive, but they can be hydroxylated under oxidative conditions which allows the CB[*n*] to be modified or grafted to a solid surface.[37] An alternative approach is to use modified glycolurils which can be used for non-hydroxyl modifications.[38] Prominent applications of functionalised CB[*n*] include supramolecular Velcro, where two surfaces, one covalently modified with CB[7] and the other with ferrocenemethylamine, allows the host–guest chemistry to be transformed into a bulk material. CB[7] has also been attached to a support bead to facilitate the capture of ferrocenelyated biomolecules. Polymeric CB nanocapsules and polymers have also been prepared from appropriately functionalised CBs to yield materials that can be modified on the surface through non-covalent interactions. The best current method to functionalise CBs appears to be the photochemical oxidation of CB[5]–CB[8],[39] which allows monohydroxylation of CB derivatives, but the products need to be separated. Challenges still remain in the search for scalable methods to produce such functionalised CBs. In a similar vein, cucurbituril analogues have been pursued as novel hosts with fine-tuned properties.

Notable milestones in CB chemistry are shown below (Table 1.1), these will be discussed and placed in context throughout the book.

Table 1.1.   Selected milestones in the field of cucurbiturils chemistry.

Year	Milestone
1905	First synthesis of cucurbituril.[6]
1981	Structural characterisation of CB[6].[8]
1983	Host–guest chemistry of CB[6] investigated[9]; first reaction inside CB[6].[40]
1986	First kinetic study on CB[6]-guest binding.[41]
1990	pH-driven CB[6]-based molecular switch.[11]
1992	DecamethylCB[5] synthesised[42]; CB[6] cation binding affinity measured.[14]
1994	First binding mechanism study.[43]
1996	Solubilisation of CB[6] in neutral media[16]; metal-directed assembly of CB[6]–polypsudorotaxanes.[44]

(Continued)

Table 1.1.   (*Continued*)

Year	Milestone
1997	CB[6]–dye complexes affinities measured.[12]
1999	First water remediation application.[45]
2000	Synthesis and isolation CB[5], CB[7] and CB[8].[19]
2001	Unpolarisabilty of CB[7] cavity investigated[46]; CB[n]-forming conditions investigated[20]; water-soluble CB*[n] (n = 5 and 6) synthesised[20]; Xe binding to CB[6] discovered[47]; first CB[8] charge-transfer complex synthesised[26]; decamethylCB[5] gas-phase complexes observed.[48]
2002	Vesicles formed from supramolecular amphiphiles[29]; CB[5]@CB[10][49]; CB[7]–viologen complexes[22] and electrochemistry[23] investigated.
2003	First direct functionalisation CB[n][37]; gas-phase CB[6]-based pseudorotaxanes observed[50]; gas sorption with decamethylCB[5].[51]
2004	Hemicucurbit[n]uril (n = 6, 12) synthesised[52]; mechanism of CB[6]-guest binding elucidated[53]; social self-sorting of CB complexes[54]; purification of rare CB[n] [55]; electrochemically activated molecular loop lock.[56]
2005	Inverted cucurbit[n]urils (n = 6 and 7) isolated[57]; CB[10] isolated[58]; enhancement of fluorescence and stability of a dye in CB[7][59]; high-affinity[31] and specificity[30] guests discovered; tryptophan peptide binding to CB[8].[60]
2006	Chiral recognition in CB[n][61]; ns-CB[10] synthesised[62]; peptide dimers in CB[8] [63]; electrochemically switchable dendrimer[64]; and pseudorotaxnes[65]; thermodynamic versus kinetic self-sorting complexes[66]; CB grafted to silica.[67]
2007	Bis ± ns-CB[6] synthesised[68]; CB[7]–guest binding affinity surpasses that of biotin-streptavaidin[32]; non-covalent immobilisation of proteins on a solid surface;[69] stimuli responsive CB[7] hydrogel;[70] CB-polymer nanocapsule synthesised[71]; label-free continuous enzyme assays developed.[72]
2008	CB[n]-formation mechanism suggested[73]; supramolecular nanovalves[74]; CB[8]-supramolecular block-copolymers[28]; CB[6]-based porous material[75]; CB[n]-functionalised nanoparticles and surfaces[76]; berberine-CB[7] binding[77]; conductive CB[7] polymer.[78]
2009	Kinetic versus thermodynamic self-sorted pseudorotaxanes.[79]
2010	Bambus[6]uril synthesised[80]; first CB[8] hydrogel synthesised[34]; CB[7]–guest activated cytotoxic nanoparticles[36]; CB[8]-based polymer synthesised[27]; toxicology of CB[n] evaluated.[81,82]
2011	CB[7]–protein binding observed[24]; CB[6]–hydrocarbon binding[83]; protein fishing with CB[7]-bead demonstrated[84]; first SERS analysis with CB[n][85]; CB[8]-mediated protein dimerisation[86]; self-sorting of pseudorotaxanes.[87]

(*Continued*)

Table 1.1.   (*Continued*)

Year	Milestone
2012	CB[6]-based hydrogels synthesised[35]; CB[8]-microcapsules formed from microdroplets[88]; "high energy water" proposed as source of high-affinity binding[89]; acyclic CBs used to solubilise drugs[90]; cooperative capture synthesis using CB[6][91]; electrochemical control of cell adhesion[92]; mono-functionalised CB[7] synthesied[38]; CB[7]–drug delivery in mice.[93]
2013	CB-based 2D polymer[94]; supramolecular Velcro[95]; 2D supramolecular organic framework synthesised[96]; control of protein activity using CB[97]; targeted drug delivery using CB[7].[98]
2014	CB[14][99]; attomolar affinity CB[7]–guest[33]; self-sorted six station pseudorotaxanes[100]; mixed metal–ligand and CB[8]–guest polymers synthesised[101]; 3D-supramolecular organic framework synthesised.[102]
2015	High affinity host–guest FRET pair[103]; photochemical hydroxylation of CB[5]-CB[8][39]; CB-based nanostructures with changeable morphology synthesised[104]; CB-controlled catalysis inside a living cell[105]; catalytic Diels–Alder reaction in CB[8].[106]
2016	Hydrocarbon binding in acyclic CBs[107]; CB[13] and CB[15] isolated[108]; first protein-based CB[7] rotaxane synthesised.[109]
2017	CB[6]-mediated bioorthogonal protein labelling[110]; high-affinity guests for CB[8][111]; catalytic Diels–Alder reaction in CB[7].[112]

Chapter 2

Cucurbiturils: Syntheses, Structures and Properties

2.1. Synthesis and Isolation of Cucurbit[*n*]urils

2.1.1. *Classical synthesis of cucurbit[6]uril*

The synthesis of cucurbit[6]uril has not changed much since Behrend first reported it in 1905. Behrend reported the acidic condensation of glycoluril in the presence of excess formaldehyde which yielded an amorphous insoluble material. The material can be dissolved in hot sulphuric acid, after dilution with water, and on further boiling a crystalline precipitate is obtained. The chemical equation showing the preparation of cucurbit[6]uril is shown in Figure 2.1.

Seventy-five years later, Mock reinvestigated the original synthesis of cucurbit[6]uril and was able to reproduce it, albeit at slightly reduced yield (40–70%) compared to that originally reported.[8] Mock and Freeman, armed with modern analytical tools, solved the structure of cucurbit[6]uril by crystallising a CB[6]–Ca^{2+} complex.[8] Although higher yields of CB[6] have been disclosed in the patent literature,[113] the method used by Behrend and Mock is still the prevalent method used today.

2.1.2. *Synthesis and isolation of the other members of the cucurbit[n]uril family*

It was not until 20 years after Mock's report that a new procedure to synthesise a mixture of different size cucurbiturils was reported by

$$6 \ C_4H_6N_4O_2 \quad + \quad 12 \ CH_2O \quad \longrightarrow \quad C_{36}H_{36}N_{24}O_{12} \quad + \quad 12 \ H_2O$$

Glycoluril   Formaldehyde   Cucurbit[6]uril

Figure 2.1.   Chemical equation for the synthesis of cucurbit[6]uril.

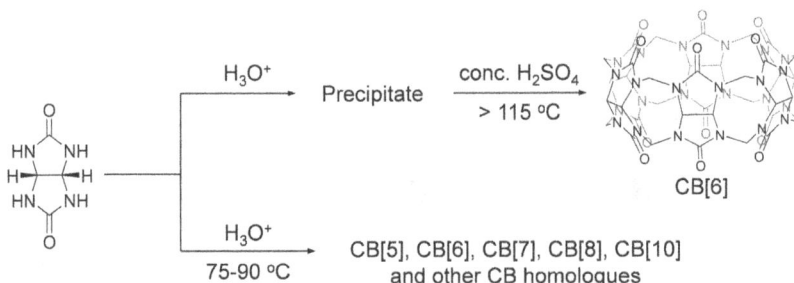

Scheme 2.1.   Synthesis of the CB[n] family.

Kim et al.[19] The new members isolated were CB[5], CB[7] and CB[8], along with CB[6]. Trace amounts of higher family members CB[9]–CB[11] were detected by mass spectrometry. The crucial difference is that the reaction is performed at a lower temperature (75–90°C, compared with 110°C in the original synthesis) allowing the isolation of the kinetic products of the reaction. It is a one-pot procedure in contrast to the original two-step method (Scheme 2.1). The content of a typical reaction mixture is 10–15% CB[5], 50–60% CB[6], 20–25% CB[7] and 10–15% CB[8]. Independently, Day et al.[20] isolated these family members and investigated the effect of conditions, such as acid type, glycoluril concentration and template effects to optimise the yield of individual CB[n]. The concentrations of glycoluril and the acid used in the reaction have a large effect on the product distribution; higher concentrations of glycoluril and concentrated HCl favour a greater proportion of larger CBs. In concentrated HCl, the ideal conversion temperature of the oligomer to the CB[n] was 80–100°C, although the conversion was possible at 50°C with extended reaction times. Although a high-yielding synthesis of CB[7] (49%) using a modified procedure has been reported,[114] typical yields of CB[7] tend to be lower (~20%).[19,20]

As well as the main cucurbituril family members, there are several other species formed under the reaction conditions. CB[10] is formed as a complex with CB[5], CB[5]@CB[10].[49] The other components of the CB reaction mixture are the inverted CB[*n*] (*i*CB[*n*], *n* = 6 or 7) family.[57] The *i*CB[*n*] are diastereomers of CB[*n*] where the methine protons on one of the glycoluril units point into the cavity rather than out; they are produced in 2.0% and 0.4% yields in the reaction mixture, respectively. The X-ray crystal structures of the CB[*n*] family members are shown in Figure 2.2.

The chemical shifts from NMR are useful for identifying the different CB[*n*]. The difference between the two sets of methylene protons increases progressively from 1.42 ppm, for CB[5], to 1.65 ppm for CB[8]. Similarly, the chemical shifts from the methylene and methine carbons move downfield with increasing ring size. In the crude reaction mixture, determination of the relative amounts of CB[*n*] is difficult due to the overlapping peaks in the ¹H NMR. ¹³C NMR is much more powerful because the peaks corresponding to each CB[*n*] are different across the family members (Table 2.1).

As the popularity of cucurbiturils has grown, more efficient synthesis methods have been sought. Microwave-heated reactions appear to be

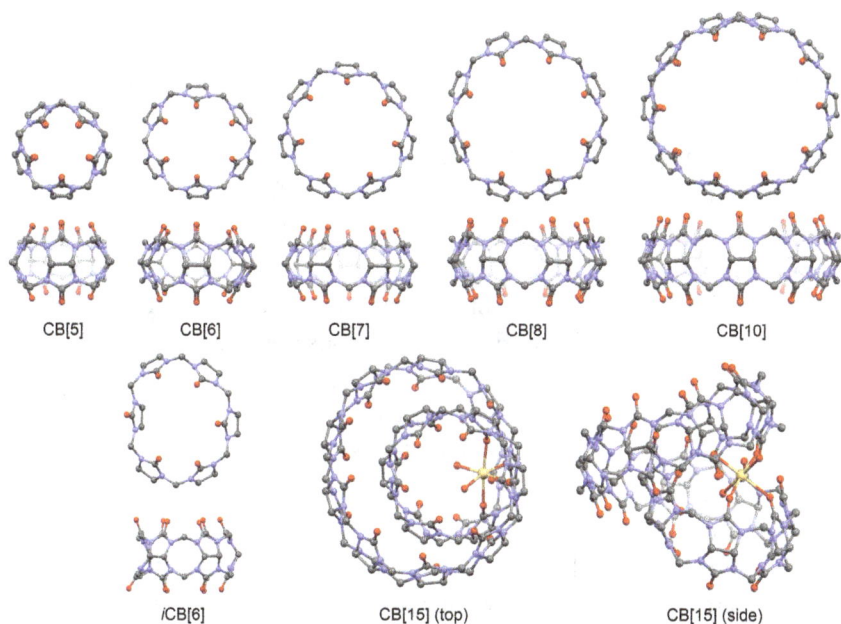

Figure 2.2.   X-ray crystal structures of CB[5]–CB[8], CB[10], *i*CB[6] and CB[15]–Cd²⁺.

Table 2.1. ^{13}C NMR chemical shifts for each member of the CB[*n*] family. (From Refs. 20, 57, 58, 99 and 108)

CB[*n*]	Methylene (ppm)	Methine (ppm)	Carbonyl (ppm)
CB[5][a]	51.8	70.6	157.9
CB[6][a]	53.2	71.7	158.0
CB[7][a]	54.4	72.7	158.3
CB[8][a]	55.4	73.6	158.6
CB[10][b]	56.9	74.6	159.2
CB[13][c]	52.5	70.5	155.9
CB[14][c]	53.2	72.0	157.1
CB[15][c]	53.1	71.9	159.3
*i*CB[6][d]	51.9, 51.7, 51.3	70.8, 70.6, 70.3, 70.1, 69.8, 62.6[e]	157.1, 166.9, 156.3, 156.0
*i*CB[7][c]	53.3, 53.1, 52.7, 52.5	71.9, 71.8, 71.7, 71.4, 71.0, 64.4[e]	157.5, 157.0, 156.9, 156.7

Notes: NMR solvents: [a]50% DCl–D$_2$O; [b]20% DCl–D$_2$O; [c]D$_2$O; [d]35% DCl–D$_2$O; [e]Inverted methine carbon.

a promising method for scalable synthesis of large amounts of CBs, more efficiently.[115–117]

An easy way to isolate CB[*n*] from the acidic reaction mixture of glycoluril and formaldehyde is achieved by the fractional crystallisation and dissolution of the reaction mixture (Scheme 2.2).[19] For example, crystals of CB[8] appear at the bottom of the reaction mixture on standing overnight. CB[6] is isolated simply by adding water to the reaction mixture to induce crystallisation. From the aqueous portion, CB[5] and CB[7] are isolated and further separated by fractional crystallisation with acetone–water and methanol–water mixtures, respectively. Effective purification is critical; so, great care should be taken to avoid contamination of the desired family member with others or organic solvent molecules since even minor levels of contamination can cause significant problems when determining binding affinities. The problem of contamination is most apparent with the water-soluble family members; for example, CB[7] frequently contains a small amount of CB[5] and *i*CB[7]. It is challenging to remove all traces of acid that are used in the synthesis of

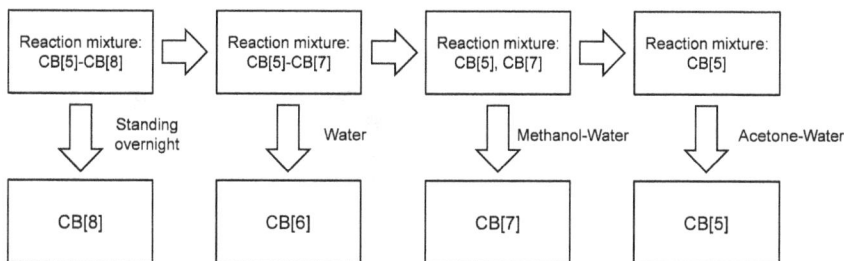

Scheme 2.2.   Representative isolation of individual CB[*n*] family members by dissolution and fractional crystallisation. In practical terms, the reaction conditions and isolation protocol are optimised for the desired family member.

CBs, and frequently methanol and acetone, depending on the isolation protocol. Free CB[6] can be obtained by performing the synthesis reaction in HCl and removing the acid in a vacuum oven.[75] It should be noted that the separation of the family members is affected by traces of acid present in the mixture as well as by NH_4^+ ion, which may be a product of thermal decomposition of glycoluril. Ammonium ions form a particularly tenacious bis-complex with CB[5] and decamethylCB[5]. Free decamethyl CB[5] can be obtained by sequential treatment with base, ion exchange and removal of the remaining NH_3 by heating under reduced pressure.[51]

CB[10] was first isolated in the form of an inclusion complex, CB[5]@CB[10][49]; guest-free CB[10] can be liberated from the CB[5]@ CB[10] by displacement of the CB[5].[58] The CB[5] guest can be replaced by a cationic guest, namely melamine diamine. The melamine is then rendered neutral by acetylation with acetic anhydride, thereby reducing the affinity of the complex. Excessive washing followed by recrystallisation affords the free CB[10].

Tao *et al.*[99,108] reported the largest CB family members to date: CB[14], and later CB[13] and CB[15]; all of these family members exhibit an unusual twisted figure of eight structure. The twisted CBs were isolated by extracting the water-soluble components (mainly CB[5] and CB[7]) from a precipitate of the standard CB reaction mixture. This fraction was subjected to ion exchange chromatography for up to 6 months, and the twisted CBs were found to be a minor component. The isolation of CB[13–15] begs the question: why are these isolatable whereas the other missing members CB[9], CB[11] and CB[12] are only seen in trace amounts? A likely explanation is that macrocyclisation step is difficult with such large oligomers. However, the twisted structure — if it exists in solution — may promote cyclisation.[118] For the other members, the entropic penalty required to cyclise may be too large. The X-ray crystal structures of the family members are shown in Figure 2.2.

Other groups have reported approaches to purify individual CB[n]. For example, an environmentally friendly method to separate the water-soluble members, CB[7] and CB[5], which are difficult to separate from each other, has been reported.[119] An alkyl imidazolium-based ionic liquid that forms a complex with CB[7] was employed; this newly formed complex then precipitates. CB[5] can then be crystallised from the aqueous phase, and free CB[7] is obtained by removal of the cationic guest with PF_6^- as the counterion and then the free CB[7] can be precipitated from methylene chloride. Chromatographic isolation of CB[7] can be achieved by using a harsh acidic mobile phase of HCO_2H (88%)–HCl (0.2 M) (1:1 v/v).[73] The separation of individual CB family members by dissolution and ion exchange chromatography on Dowex has also been reported. In this procedure, the reaction mixture is first separated into water soluble and insoluble components. The water-soluble components are loaded onto a Dowex column and are eluted by acid; their elution order is *i*CB[7], CB[5], CB[7], and much later CB[13]–CB[15].[108]

As mentioned above, the CB[n] often carry water, acid molecules and other cations, which are very difficult to remove completely. To assess the purity of individual CB[n], elemental analysis or titrations with specific guest molecules must be performed. For example, the Kaifer group use a UV-vis titration to measure the amount of cobaltocenium hexafluorophosphate that was encapsulated in samples of CB[7] and CB[8].[120] The Masson group assess the purity by employing ¹H NMR titrations with 1,6-hexadiammonium, *p*-xyelendiammonium or 1-adamantylpyridinium cations for CB[6], CB[7] and CB[8], respectively.[7] The titration methods are particularly useful because they measure how much CB[n] is available. The Kim group use two molecules to quantify the amount of CB[7] and CB[8] available. They use maleic acid as an internal standard, which does not interact with CB[n], and *p*-xyelendiammonium dihydrochloride to aid solubility and bind to CB[n] to quantify how much CB[n] is available.

Each lab that works on CB chemistry has their own procedures to synthesise and isolate the desired family member or derivative, but the yields, purity and product distributions do not always reproduce well in other labs. A challenge for the future of CB chemistry is to agree on unified synthetic procedures that specify the purity and supplier of starting materials, reagents and crystallisation solvents, as well as detailed procedures.

2.1.3. *Mechanism of cucurbit[n]uril formation*

The mechanism of cucurbituril formation has been investigated in detail by the Isaacs group. They realised that CB formation is likely a stepwise elongation process with the macrocycle closure as the final step (Figure 2.3). Their strategy was to perform the classical CB synthesis in formaldehyde-deficient conditions to isolate intermediates in the CB-formation pathway.

Informed by earlier studies on the dimerisation of glycolurils,[121–123] each acyclic oligomer (dimer to hexamer) was isolated by controlling the amount of formaldehyde in the reaction mixture, thereby showing a stepwise elongation process.[73] Once the chain becomes long enough to reach its tail, i.e. a pentamer or hexamer, then ring closure to the cucurbituril can be completed by the addition of two further equivalents of formaldehyde. The ring closure proceeds via the addition of the first equivalent of formaldehyde to yield a hydroxylated intermediate, which then rearranges to *ns*-CB[6] (*ns*-CB[6]), an isolatable product. The second equivalent of formaldehyde undergoes condensation to complete the CB[6]. *i*CBs are also formed, as kinetic products, in the reaction mixture, and their mechanism of formation and conversion to CB has been investigated.[57,124]

These mechanistic insights informed studies on the synthesis of acyclic analogues and the *ns*-CB family. The reader is directed to reviews that give more details on the mechanistic studies.[125,126]

2.2. Structure and Properties of the Cucurbit[*n*]uril Family

2.2.1. *Structural features of the cucurbit[n]uril family*

The cucurbit[*n*]uril family (CB[5]–CB[10]) consists of highly symmetric, cyclic methylene-bridged glycoluril oligomers with a hydrophilic rim on

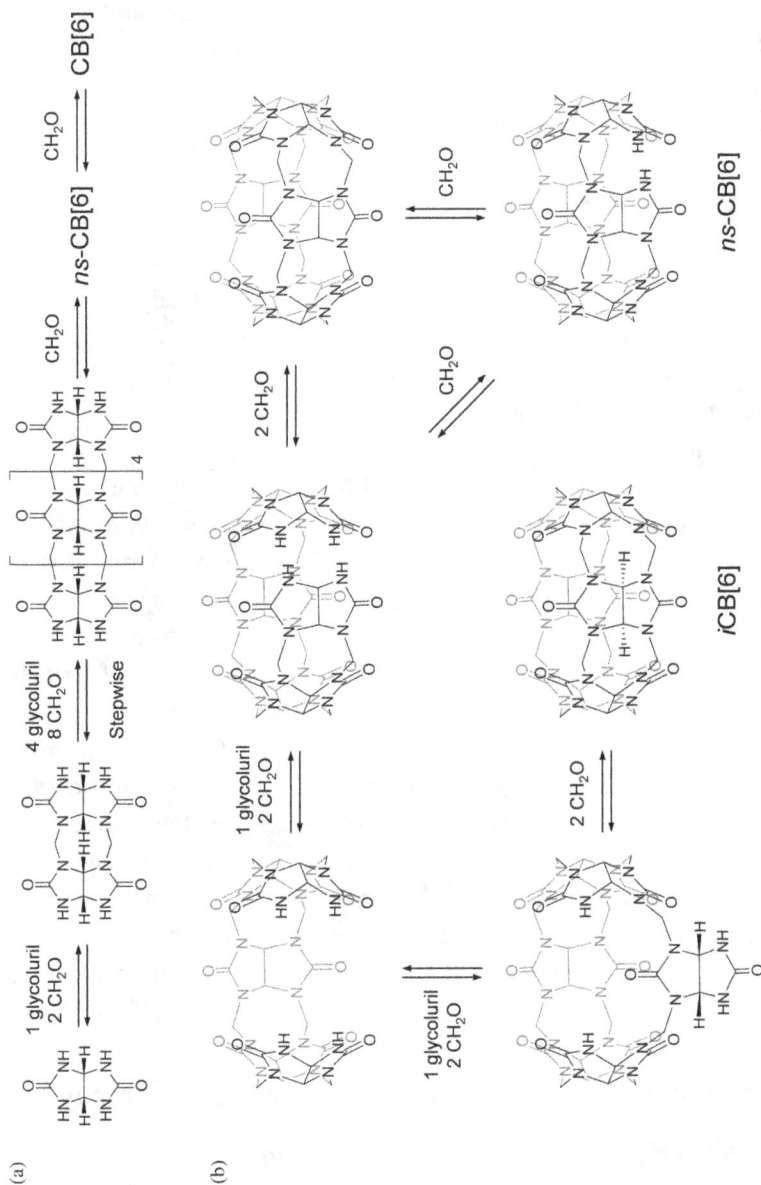

Figure 2.3.   (a) Stepwise elongation of glycolurils and cyclisation to CB[6] via *ns*-CB[6]. (b) Various cyclisation mechanisms to give *i*CB[6], *ns*-CB[6] and CB[6].

Table 2.2.   Structural parameters (in Å) and physical properties of the CB[*n*] family.

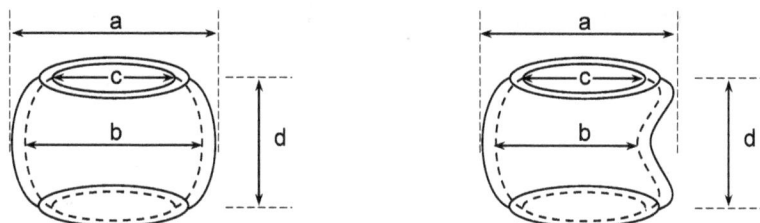

	CB[5]	CB[6]	CB[7]	CB[8]	CB[10]	*i*CB[6]	*i*CB[7]
Outer diameter (a)	13.1	14.4	16.0	17.5	20.0	10.7–14.4	11.2–16.0
Inner cavity (b)	4.4	5.8	7.3	8.8	11.7	3.8–5.8	5.4–7.3
Portal diameter (c)	2.4	3.9	5.4	6.9	10.0	4.3–3.9	6.7–5.4
Height (d)	9.1	9.1	—	9.1	9.1	9.1	9.1
Volume/Å^3	82	164	279	479	870	—	—
Water solubility/mM	20–30	0.018	20–30	<0.01	—	—	—
Stability/°C	>420	425	370	>420	—	—	—

each side comprised of urea carbonyls (termed the portal) and a hydrophobic interior (termed the cavity, see Figure 2.2). Each member of the family shares a common depth of 9.1 Å.[21] The other structural dimensions of CB[5]–CB[8]: their equatorial widths, angular widths and volumes increase systematically with ring size, and the cavity diameter starts at 4.4 Å and increases to 8.8 Å (Table 2.2).[21] The portals are the entrance to the hydrophobic cavity, and they are ~2 Å narrower than the cavity itself; this provides a steric barrier to guest association and dissociation. The flexibility of the CBs increases with sizes, which means they can become distorted by guest inclusion or packing into a crystal. Consequently, the crystal structure of CB[10] is quite ellipsoidal with transverse and conjugate diameters of 11.3 and 12.4 Å, respectively.[126] In the case of *i*CB[*n*], since one of the glycolurils points inwards the volume of the cavity is reduced. The twisted CB[*n*] are less voluminous than the smaller members, in fact they behave like two connected CBs and arrange themselves into a shell-like structure.[127]

2.2.2. *Physical properties of cucurbit[n]uril*

A prominent feature of the cucurbituril family is their poor solubility in water and organic solvents. CB[6], CB[8] and CB[10] are essentially insoluble in water, while CB[5] and CB[7] are modestly soluble (2–3 × 10^{-2} M^{-1}). CB[13–15] are soluble in water and DMSO.[99] The differences in water solubility across the family members are a consequence of their crystal structures. CB[6] and CB[8] generally form a well-packed honeycomb-like structure where there is little room for water to solvate the CB. Consequently, CB[6] and CB[8] remain crystalline upon drying, whereas CB[5] and CB[7] become amorphous. These structural observations explain why CB[6] and CB[8] are insoluble in water, yet CB[5] and CB[7] have modest water solubility.[128] The portals of CB[n] are weak bases: The pK_a of the conjugate acid of CB[6] is 3.02.[14] Consequently, the solubility of cucurbiturils improves markedly in acidic media such as 50% aqueous formic acid.[8] Kim *et al.*[16] found that CB[6] dissolves in aqueous solutions that contain sodium ions paving the way for the measurement of binding affinities in neutral water. The origin of the solubility increase is the complexation of sodium cations to the portal of the CB. Modification and derivatisation of CB[n]s to improve their solubility has been the subject of intense research and is discussed in later sections.

The CB[n] family have exceptional thermostability. In the cases of CB[5], CB[6] and CB[8], no decomposition is observed up to 420°C; CB[7] begins to decompose at 370°C. CB[8] decomposes to smaller family members during prolonged heating at 100°C in HCl; the smaller CB family members do not decompose under the same conditions.[20]

The properties of the CB portal and cavity play a crucial role in molecular recognition. The electrostatic potential of the portal region of CB[n] is strongly negative and contrasts starkly with the non-polarised interior (Figure 2.4).[129] This figure shows the calculated electrostatic potential of CB[7] compared with that of β-cyclodextrin.[21] The hydrophobic cavity of CB[n] is a remarkably non-polar and unpolarisable environment. The seminal work in this area was conducted by Nau *et al.*[46] and they determined the polarisability of the cavity using the dye DBO (2,3-diazabicyclo[2.2.2]oct-2-ene) as a spectrophotometric probe. The polarisability (*P*) has a value of 0.12 (*c.f. P* = 0.159 for perfluorohexane, the least polarisable solvent known and *P* = 0.0 for the gas phase).

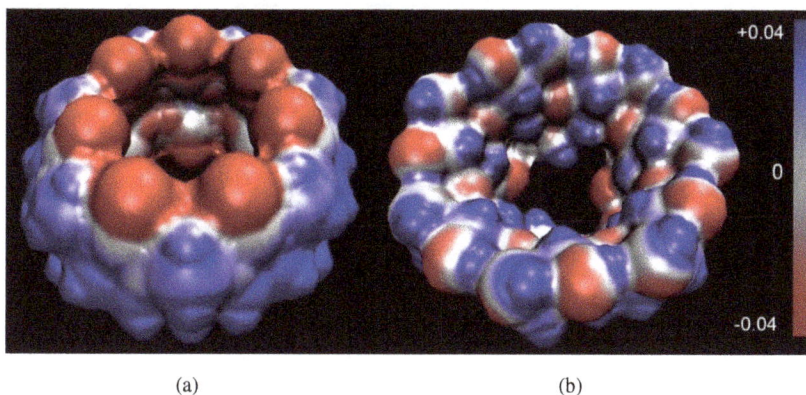

(a)   (b)

Figure 2.4.   Electrostatic potential of (a) CB[7] and (b) β-cyclodextrin; this plot shows the highly negatively polarised portal of CB[7]. (Modified from Ref. 21)

This study showed that the cavity of CB[7] was extremely unpolarisable. Other studies, one utilising fluorinated guests and ^{19}F NMR,[130] and others using dye encapsulation experiments and fluorescence measurements[59,131] have provided further evidence that the CB[7] cavity behaves more like the gas phase than any other solvent.

These observations can be explained by considering the peculiar microenvironment of the CB[n] cavity, which is vacuum-like in some ways.[132,133] There are no bonds, functional groups or lone pairs of electrons pointing into the interior of a CB[n]; the macrocycles are not aromatic and the lone pairs on the nitrogens are delocalised around the carbonyls.

2.3. Synthesis of Cucurbit[n]uril Derivatives

2.3.1. *Cucurbituril derivatives from modified building blocks*

Glycoluril and formaldehyde are the building blocks of cucurbituril. The replacement of glycoluril with substituted glycolurils or formaldehyde with another aldehyde should, therefore, lead to the preparation of substituted cucurbiturils. Two strategies have been applied to achieve this (Scheme 2.3). The first is simply to prepare glycoluril derivatives and react those with formaldehyde, either alone or as a mix-and-match with glycoluril. The second strategy is to prepare linear acyclic CBs and close

(a)

R =Me, Ph, -C$_3$H$_6$-, -C$_2$H$_4$-

(b)

X = OH, NO$_2$
or CO$_2$H

CB[6] derivative

mono-substituted CB[7]

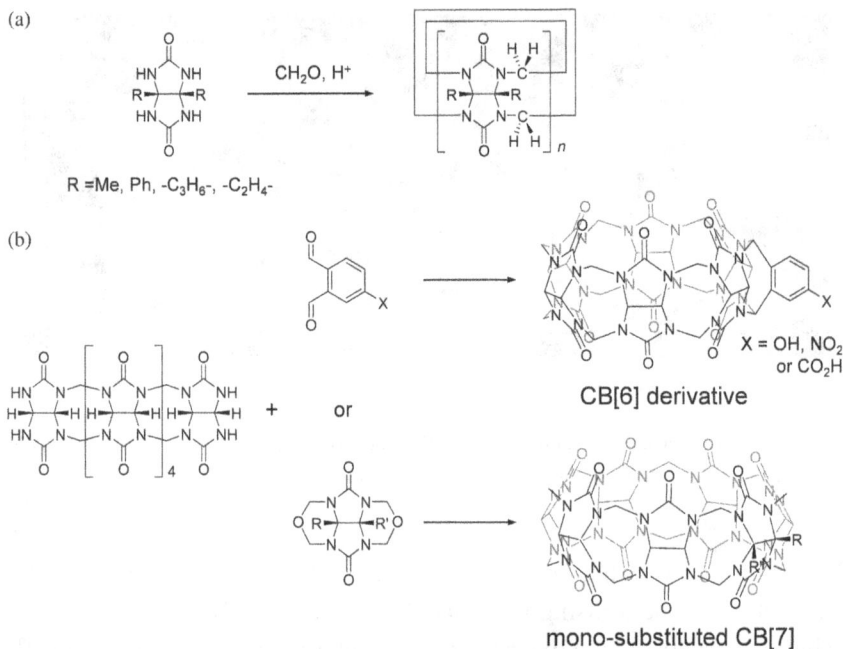

Scheme 2.3.   Preparation of functionalised CB[n] by (a) modified glycolurils and (b) by formaldehyde surrogates.

the macrocycle with a modified aldehyde or aldehyde surrogate. These strategies will be discussed in turn; representative structures synthesised using these methods are shown in Figure 2.5.

Glycoluril is the product of an acid-catalysed condensation reaction between urea and glyoxal. The replacement of glyoxal with various 2,3-butadiones leads to preparation of glycolurils substituted at the methine positions (Scheme 2.3(a)). Dimethylglycoluril was the first alkyl-substituted glycoluril used to prepare a substituted CB[n]. In 1992, Stoddard et al.[42] reported the discovery of decamethylCB[5]. Interestingly, this discovery came 8 years before the synthesis and isolation of CB[5], CB[7] and CB[8]. X-ray analysis reveals that Me$_{10}$CB[5] has an almost identical portal size and cavity volume to CB[5]; the methyl groups can be seen protruding from the methine positions of the CB. However, like CB[5], it is quite insoluble in most common solvents. A small amount of permethyl-substituted CB[6] can also be formed under these conditions.[55]

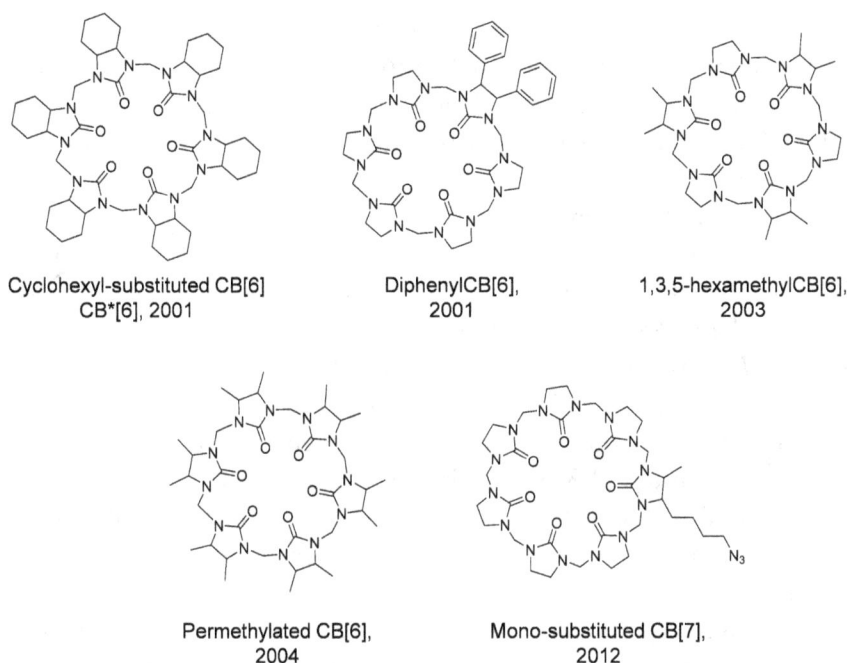

| Cyclohexyl-substituted CB[6] CB*[6], 2001 | DiphenylCB[6], 2001 | 1,3,5-hexamethylCB[6], 2003 |

| Permethylated CB[6], 2004 | Mono-substituted CB[7], 2012 |

Figure 2.5.   Structures of various substituted cucurbiturils, viewed from above.

Shortly after the synthesis of the CB[5]–CB[8] had been realised, Kim *et al.*[134] prepared cyclohexyl-substituted CB[5] and CB[6], termed CB*[5] and CB*[6] (Figure 2.5). The precursor glycolurils were prepared by the condensation of cyclohexane-2,3-butadione with urea. The substituted glycoluril was then subjected to cucurbituril formation conditions. Again, the dimensions of the modified CBs were almost the same as for the parent CBs. The fused cyclohexyl units decorate the outside of the "equator" of the CB. The most remarkable property of these derivatives is their surprising solubility in water: it is comparable to that of α-cyclodextrin, meaning that binding experiments could be performed in *neutral* water. Furthermore, they were soluble in common organic solvents. In contrast to the native CB[*n*], the reaction mixture distribution of substituted CB*[*n*] tends to be weighted towards CB*[5] and CB*[6]; substituted CB*[7] and CB*[8] only exist in trace amounts. A similar trend is seen with the analogous cyclopentyl[135] and permethylated CBs.[55] The first phenyl-substituted derivative was reported in 2002.[136] This was also the first asymmetrically functionalised CB[6] which was prepared by mixing five equivalents of glycoluril with one equivalent of diphenyl glycoluril to afford diphenylCB[6]. The mixed glycoluril strategy has also been used to prepare partially cyclopentyl-substituted CB[6] derivatives by varying the stoichiometry of the components.[137]

Besides solubility, the other motivation for derivatising cucurbiturils is to incorporate reactive groups that are synthetically tractable so that CBs can be chemically immobilised or conjugated to other molecules. Kim et al.[138] disclosed in a patent the synthesis of di(aminophenyl)CB[n] by using a mixed building block strategy. They synthesised 3 and 4-dinitrophenylglycoluril and condensed it together with five equivalents of glycoluril to afford 3 or 4-dinitrophenylCB[6]. The nitro group was then reduced with either $Sn^{2+}–Cl_2 \cdot HCl$ or with ammonium sulphide to afford the amine, which can be further functionalised. Isaacs et al. prepared a different monoamine, or carboxylic acid functionalised CB[6];[139] they closed an acyclic glycoluril hexamer with either 3-nitro or 4-carboxylic acid-phthalaldehyde (Scheme 2.3(b)) to give an amine- or carboxylic acid-functionalised CB[6]. They later extended this approach by replacing the dialdehyde with 4-hydroxy phthalaldehyde to give a hydroxyl-functionalised CB.[140]

When substituted glycolurils react with formaldehyde under acidic conditions, dimethyl-substituted cyclic diethers can be isolated.[141] These cyclic diethers can be used as building blocks in the synthesis of CB[n] derivatives (Scheme 2.3(b)). In 2003, Day et al.[142] introduced a method that afforded partially methyl-substituted CB[n] (n = 5–7), where a dimethyl-substituted cyclic diether was condensed with glycolurils. This approach yielded 1,3,5-hexamethylCB[6] as the major product. Similarly, 1,4-tetramethylCB[6] can be obtained by reaction of the dimethyl-substituted cyclic diether with a glycoluril dimer in a 2:1 ratio.[143] Substituted cyclic diethers have also been used in the synthesis of CB[7] derivatives. In this case, the condensation of an acyclic glycoluril hexamer was carried out in the presence of various substituted cyclic diethers to afford substituted derivatives, which can be further functionalised for useful applications.[38] Typically, their substituted glycoluril derivative has an appended alkyl chloride, which can be further modified to prepare CB[7] conjugates such as N_3 or biotin conjugated-CB[7] (Figure 2.5).[98] A similar method was used to prepare disulphonate CB[7] derivatives[144] with the aim of improving CB[7]'s solubility in water.

2.3.2. *Direct functionalisation of cucurbiturils*

The low solubility of the cucurbituril family (CB[6] and CB[8] in particular) in water or organic solvents, and the lack of functionalisation methods for the CB skeleton were the major limitations at the early stage of cucurbituril research. While the addition of alkyl or phenyl units to the skeleton by incorporation of functionalised building blocks improved the solubility to an extent, these building block methods often lead to a mixture of products that are very challenging to separate. A more pleasing solution to this problem is to functionalise the CB[n] of interest directly.

Direct functionalisation of cucurbiturils is challenging because CB[n] are thermally and chemically stable, even under forcing conditions. Examination of the molecular structure shows that there are a few obvious functionalisation points. In 2003, Kim *et al.*[37] discovered that hydroxyl-functionalised CBs could be prepared by reaction of a CB with $K_2S_2O_8$ in water (Scheme 2.4). For the smaller family members, CB[6] and CB[5], the yield of perhydroxyCB[n] was 45% and 42%, respectively. For the larger CB[7] and CB[8], the yields were ~5%, presumably because of the instability of the respective perhydroxylated products. Later, careful optimisation of the reaction conditions allowed the synthesis of mono-functionalised CB[7].[95] Six equivalents of K_2SO_4 and a sub-stoichiometric amount of $K_2S_2O_8$ favours the formation of the monohydroxylated product. Using another oxidative approach, Scherman *et al.*[45] were able to synthesise monohydroxyCB[6]. The oxidant, in this case, was ammonium persulphate and the reaction was mediated by a guest imidazolium molecule, which allowed the CB[6] to be dissolved in water. MonohydroxyCB[8] was synthesised for the first time in 2015 by the groups of Ouari and Bardelang.[39] They reported an oxidative method that employs H_2O_2 and UV light to generate an OH radical; the radical preferentially extracts the methine proton, from the CB skeleton, to yield a tertiary radical, which then affords the hydroxylated CB[n] upon radical recombination with another OH radical. This method can be used with all

R = H or OH, n = 5-7

(a)

CB[5]-CB[8]   MonohydroxyCB[5]-CB[8]

(b)

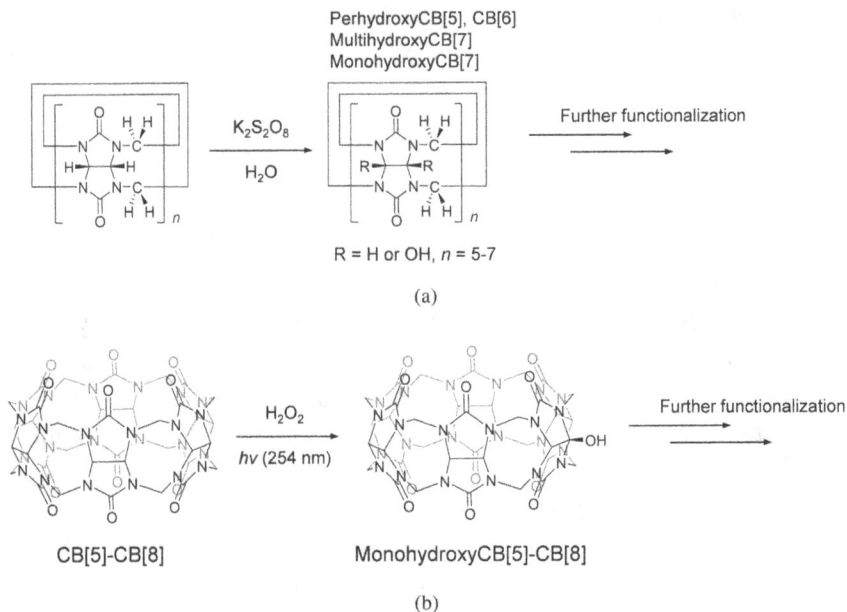

Scheme 2.4.   (a) Direct oxidative functionalisation with potassium persulphate and (b) by photochemical oxidation.

the main members CB family (CB[5]–CB[8]); however, careful chromatographic analysis revealed the presence of non-functionalised and di-hydroxylated CBs as well.[146] Nevertheless, this remains an attractive salt-free method for the production of monohydroxylated CBs.

Hydroxylation of the cucurbituril skeleton improves the solubility of cucurbiturils without affecting the dimensions of the CB structure. X-ray analysis of the perhydroxyCB[5] and CB[6] shows the hydroxyl groups on the periphery of the CB skeleton and the portal dimensions and the cavity size are unaffected.[37] Since the hydroxy derivatives are soluble in DMF and DMSO further chemistry can be performed on them. Acetylation of the hydroxy groups with anhydrides further increases the organic solubility of these derivatives. Furthermore, the hydroxyls are a reactive handle through which further chemistry can be performed. A particularly useful transformation is the allylation of the hydroxy(s) with allyl bromide to yield (per)allyloxyCB[n], which can then be functionalised with a thiol-ene photo-addition reaction or by metathesis. Several CB-based materials have been prepared by following this route, and they will be discussed in Chapter 6.

2.4. Cucurbituril Analogues and Related Compounds

There have been several analogues of cucurbiturils prepared including acyclic cucurbiturils and those made from glycoluril analogues. Representative structures are shown in Figure 2.6 and their preparations are discussed in this section.

2.4.1. *Acyclic cucurbituril analogues*

Isaacs *et al.* has pursued the development of acyclic cucurbituril analogues with the aim of making CB-like hosts with tunable properties. The

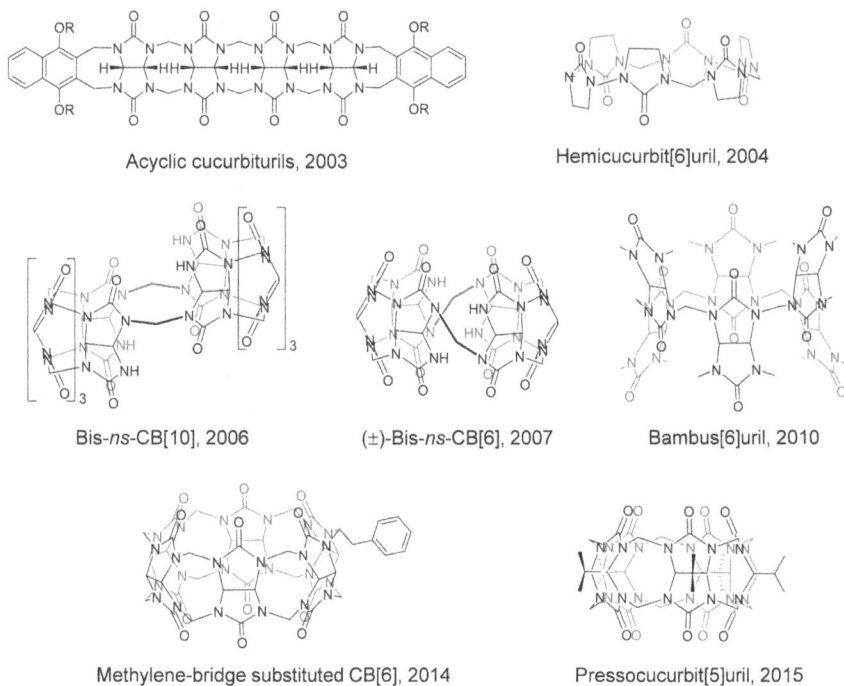

Acyclic cucurbiturils, 2003

Hemicucurbit[6]uril, 2004

Bis-*ns*-CB[10], 2006

(±)-Bis-*ns*-CB[6], 2007

Bambus[6]uril, 2010

Methylene-bridge substituted CB[6], 2014

Pressocucurbit[5]uril, 2015

Figure 2.6. Representative structures of CB analogues and related acyclic cucurbiturils analogues.

idea was that sufficiently preorganised acyclic oligoglycolurils will have CB-like binding behaviour, but will also have expanded host–guest chemistry because of the structural flexibility and greater synthetic scope to tune the properties.

The simplest acyclic analogue is based on a glycoluril dimer. In this case xylene-capped diglycolurils were used, but these were not methylene-bridged; instead, they were bridged by durene (1,2,4,5-tetramethylbenzidine), thereby increasing the number of diastereomeric positions from two to four.[147] These diastereomeric positions can either be *syn* or *anti*; the methine substituent dictates the diastereomeric preference of the reaction. *Anti* is the thermodynamic product because it reduces steric clashes between the methine substituents. When all four positions are *anti*, then the molecule is in cucurbituril-like conformation: it is an extended C-shape, with all of the methine substituents pointing outward. This conformation also minimises self-associations. These acyclic analogues are similar in size to CB[6] and retain much of the host–guest chemistry, albeit at reduced affinity (around 180-fold reduction for some alkylammonium guests). In context, this reduction is quite small in comparison to other host molecules; indeed, when other macrocycles become acyclic they lose much more of their binding ability. This is probably due to the high degree of preoganisation instilled by the diastereomeric preference of the oligomers.

The main reason for the loss of binding affinity seems to be the electrostatic potential of the portal being compromised by the addition of non-glycoluril groups. More native-like acyclic cucurbiturils needed to be synthesised to address this issue. To this end, Sindelar *et al.*[148] reported xylene-capped glycoluril trimers. Their approach was to prepare a mono-xylene-capped glycoluril monomer and condense it with 0.5 equivalents of a methyl-substituted glycoluril diether to afford the desired trimer. The trimer could also be capped with propylene, on both sides, leading to a more open C-shaped structure. Isaacs *et al.*[49] continued along this line to prepare xylene-capped tetramers. They prepared a glycoluril dimer and then reacted that with glycoluril diether to elongate by one unit on each side. The oligomers were then capped with *o*-substituted xylenes. The hydroxyls on the xylene moiety allow further functionalisation. For example, they have been converted to carboxylates to enhance solubility.

In later work, the xylene groups were decorated with sulphonate groups to make extremely water-soluble hosts (Figure 2.6).[90,150]

2.4.2. Nor-seco-cucurbiturils

The *nor-seco*-cucurbituril family (*ns*-CB[*n*]) was introduced in the previous section as an intermediate in the formation of CB[6]. Their characteristic feature is that they are missing one or more methylene bridges in their structure. As shown in the mechanistic studies above, they can be synthesised under formaldehyde-deficient conditions.

The simplest of the *ns*-CB[*n*] family is *ns*-CB[6], where the penultimate methylene bridge forms in its normal position, but there is no formaldehyde left to form the final bridge. The remaining amines present an opportunity to modify the ring with another aldehyde. For example, *ns*-CB[6] can be treated with *o*-phthalaldehyde to close the structure.[151] This modification seems to widen the portal where the modification was made and increased the electrostatic potential at the opposite portal.

The first chiral member of the cucurbituril family to be discovered was (±)-bis-*ns*-CB[6] (Figure 2.6).[68] It forms when two glycoluril trimers form a methylene bridge diagonally with each other, i.e. an NH from the top of the first trimer condenses with a formaldehyde onto an NH on the bottom of the other trimer. The process is repeated at the other side to complete the macrocycle. It is similar in size and shape to CB[6]. Its most interesting feature is that it can distinguish chiral guests and form diastereomeric complexes.

The largest member of the *ns*-CB family is bis-*ns*-CB[10] (Figure 2.6).[126] Bis-*ns*-CB[10] is formed in a similar fashion to bis-*ns*-CB[6], except it occurs between two pentamers rather than trimers. Bis-*ns*-CB[10] has two equivalent cavities that are slightly offset and connected by two methylene bridges. The two cavities allow bis-*ns*-CB[10] to form either 1:2 host:guest complexes with small guests or 1:1 complexes with larger guests.

2.4.3. Hemicucurbit[n]urils

Hemicucurbiturils are *N*-methylene-linked macrocycles that are composed of ethyleneurea units, rather than glycolurils. They are prepared by the condensation of ethyleneurea with formaldehyde in acid. They resemble a CB that has been split across the equator; the carbonyl group can be facing up or down because the links between the ethyleneurea units are free to rotate. The most stable structure is the one where carbonyl orientation alternates between up and down at each unit.

Hemicucurbiturils were first reported by Miyahara *et al.*[52] and they come in two sizes: $n = 6$ or 12 (Figure 2.6). The ring size can be easily, and efficiently, dictated by changing the temperature and the acidity of the reaction. Unlike CBs, *h*CBs are soluble in some common solvents.

Another class of *h*CB[*n*] are prepared from cyclohexylureas. The cyclohexylurea is a chiral molecule and when used in enantiopure form gives enantiopure hemicucurbiturils.[152] These come in two sizes, either *cych*CB[6] or *cych*CB[8], which can be determined with an anionic template.[153]

2.4.4. *Bambus[6]uril*

A relative of the cucurbituril family are the bambus[*n*]urils, and these were introduced by Sindelar *et al.*[80] (Figure 2.6). They are neutral macrocycles structurally similar to cucurbiturils, but they have only one methylene bridge between the glycolurils; the unbridged positions are methylated. They are synthesised using formaldehyde and acid with asymmetric glycoluril derivatives where the "top" urea is free, and "bottom" urea is methylated, or *vice versa*. The most striking feature of bambus[6]uril is the alternating bridging position; this leads to a much deeper cavity size than cucurbit[6]uril (12.7 Å compared to 9.1 Å). The long cylindrical structure earned it the name bambusuril, after the bamboo plant. Similar to cucurbiturils, they are insoluble in most common solvents. Other members of the bambusuril family replace the methyl group with other substituents to increase their solubility.[154] Their principal use to date has been as receptors for anions. Water-soluble bambus[6]urils that are hosts for a diverse range of anions in water have been reported.[155] A closely related structure is biotin[6]uril, which is prepared by the acidic condensation of biotin in formaldehyde.[156] Like bambus[6]uril, this biotin-based macrocycle also binds anions though C–H interactions.

2.4.5. *Other analogues*

Diethylester-substituted glycoluril, in combination with cyclic diethers of glycoluril, can be oligomerised in a controlled manner, to yield dimers, trimers and tetramers.[157] The esters can be transformed to imides or acids by simple organic reactions. Isaacs *et al.*[157] found that these oligomers can be condensed with bis(phthalhydrazide) to afford analogues of CB[5], CB[6] and CB[7] that have one pair of methylene bridges replaced with a phthalhydrazide moiety. Furthermore, the periphery of the CB could be decorated with esters, imides or acids depending on the oligomer used. The phthalhydrazide group is intrinsically fluorescent and can be used to monitor guest binding events.[158,159]

Another cucurbituril analogue introduced by Sindelar *et al.*[160] is called pressocucurbit[5]uril (Figure 2.6). It is synthesised from propanediurea and formaldehyde under acidic conditions to afford a perdimethylated cyclic pentamer. The portal diameter and depth are slightly reduced, and the internal diameter is increased. However, the internal volume remains very similar to the other CB[5] macrocycles. The CB[4] version of this analogue has also been prepared, which is the smallest member of the extended CB family.[161]

CB[6] derivatives that are substituted on one of the methylene bridges have been successfully prepared. Sindelar *et al.*[162] demonstrated that mono-substituted CB[6] could be prepared by using mixtures of glycoluril, formaldehyde and an aldehyde substituted with either an alkyl, vinyl benzene or benzyl group (Figure 2.6). The substituent protrudes outwards from the CB. There are few examples of such modifications because the diastereomeric preference of the initial glycoluril dimerisation reaction is inverted by aldehydes other than formaldehyde. The chain elongation process does not occur with this diastereoisomer.[163]

Chapter 3

Host–Guest Chemistry of Cucurbit[*n*]urils

3.1. General Host–Guest Chemistry of the Cucurbituril Family

CB[*n*] possesses several structural features that allow them to form highly stable host–guest complexes. Namely, the adjacent extremes of a highly polarised portal and an unpolarised cavity, which facilitate ion–dipole and hydrophobic interactions, respectively (Figure 3.1). Another consideration is the size and shape complementarity between host and guest which maximise the van der Waals interactions between the cavity wall and the guest. The rigidity of the host molecule also helps stabilise the complex.

The electronegative portal of the CB[*n*] is highly attractive for cations. In the previous chapter, a plot of the electrostatic potential of CB[7] shows the negative polarisation of the portal (Section 2.3). The carbonyls of the portal are strongly polarised and therefore interact with various metal and organic cations. The portal diameter is shorter than that of the cavity, and this provides a kinetic barrier to guest association and dissociation. The cavity is extremely hydrophobic because there are no functional groups or lone pairs of electrons pointing into the cavity. Therefore, hydrophobic guests prefer to occupy the cavity.

While CBs have a superficial resemblance to other hosts such as cyclodextrins and calixarenes, their symmetrical structure and the polarised portal distinguish CBs from these other hosts. The major manifestation of these differences is in the host–guest chemistry, CBs discriminate

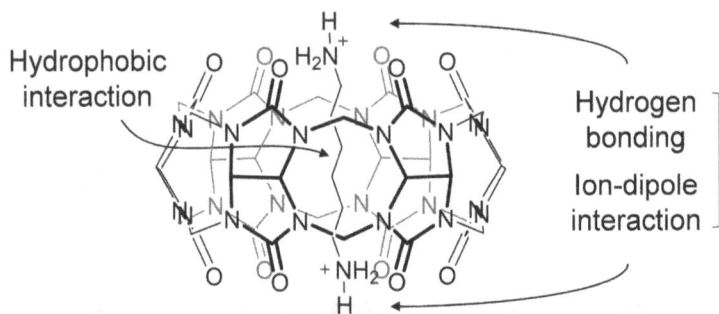

Figure 3.1.   The host–guest interactions between CB[6] and guest molecules.

on the basis of charge, whereas the other hosts do not; there are two identical portals that allow entry to the cavity (Figure 3.1), in contrast to the other hosts which are disymmetric.

The volume of the cavity increases systematically through CB[5]–CB[8] and CB[10] and this is manifested in the well-defined host–guest chemistry of each CB family member (Figure 3.2). For example, CB[5] only encapsulates small gas molecules and makes exclusion complexes with metal cations; CB[6] accommodates alkyl ammoniums and small cyclic molecules; CB[7] encapsulates larger and bulkier molecules such as metallocenes; CB[8] forms ternary complexes (two guests in the cavity) in many cases; CB[10] encapsulates porphyrins and other macrocycles.

Most basic guests experience a pK_a shift upon encapsulation inside CB[n] due to the portal-induced stabilisation of cationic guests. In particular, some dye molecules[164–165] and drug molecules[166] experience several units of pK_a shift upon encapsulation. The portal to stabilise the protonated guest is through ion–dipole interactions. With CB[6], remarkable pK_a shifts of up to 4.5 have been observed,[167] similar to that observed with enzyme–substrate interactions. With CB[7], some dye molecules exhibit pK_a shifts of 2–4 units.[164,165,168] Drug molecules can be both activated and stabilised by CB[7] because of the host-induced pK_a shift of up to 5 units.[166]

Ion–dipole interactions, hydrophobic effects and size complementarity all need to be considered when examining CB host–guest complexes. The molecules with the highest affinities tend to be appropriately sized amphiphilic molecules with a cationic group and are preorganised to make

CB[5]	CB[6]	CB[7]	CB[8]	CB[10]
N_2, O_2, Ar	Alkali metal ions			
NH_4^+	Xe, Ar, CO_2			
Alkali metal ions	THF, Benzene			
Pb^{2+}	$H_3N^+(CH_2)_nNH^{3+}$			
	$H_3N^+C_3H_6H_2N^+(CH_2)_6NH_3^+$			
	$H_3N^+C_3H_6H_2N^+(CH_2)_6NH_2^+C_3H_6NH_3^+$			

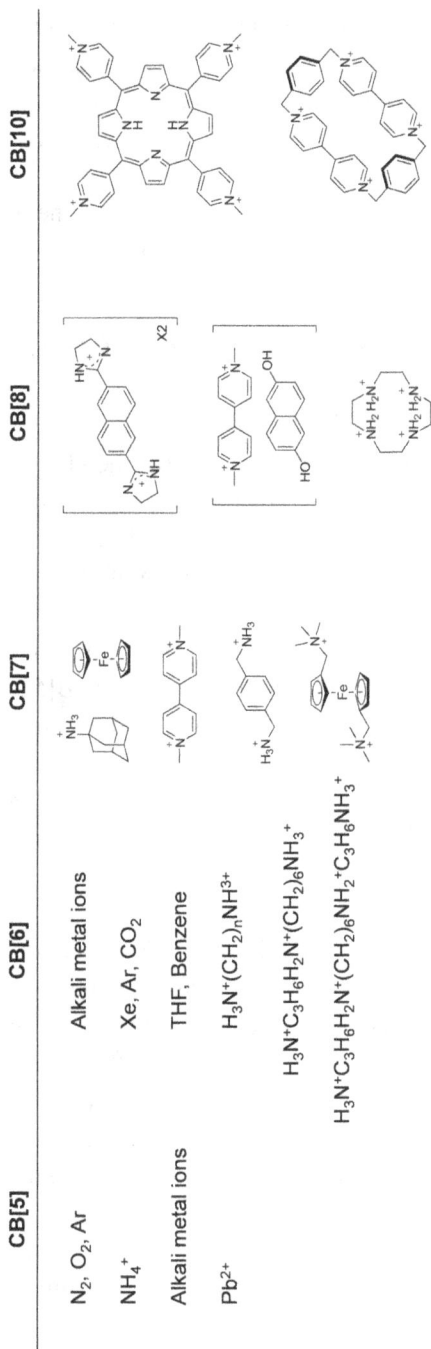

Figure 3.2. Representative guests of CB[*n*].

interactions with the CB. Another important consideration throughout this section is the media that the binding studies are measured in because this has a big influence on the binding constants. Early studies, especially on CB[6], were performed in a 1:1 mixture of formic acid and water; later, it was discovered that CB[6] could be dissolved in salt solutions or pH-controlled buffers. Binding studies done in different media are not directly comparable and should be treated with caution.

3.2. Host–Guest Chemistry of Individual CB[*n*]

3.2.1. *Cucurbit[5]uril*

Cucurbit[5]uril (CB[5]) is the smallest known member of the cucurbituril family. It has a narrow portal diameter (2.4 Å) and small cavity volume (82 Å^3). The host–guest chemistry is therefore limited to the encapsulation of small gaseous molecules and complexation of cations to the exterior of the polarised portal.

Both CB[5] and decamethylCB[5] form portal complexes with various cationic species. CB[5] and decamethylCB[5] can form exclusion complexes with alkali, alkaline earth and ammonium ions,[169] and CB[5] forms complexes with larger multicharged metals such as Fe^{2+}, Co^{2+} and others.[170] In particular, decamethylCB[5] has a remarkably high affinity for Pb^{2+}, whereas CB[5] itself does not have such high selectivity.[170]

DecamethylCB[5] can encapsulate small gaseous molecules into its cavity both in solution and in solid form.[51] DecamethylCB[5] shows size selectivity in encapsulating gases. For example, in the solid state, it easily encapsulates N_2, O_2, Ar, N_2O, CO and CO_2, and they can be easily released by heating. Smaller gases move in and out of the cavity too easily to be encapsulated. Larger gases such as Kr, Xe and CH_4 are not readily encapsulated, at least in the solid state. In solution, all of these gases can be encapsulated, although the solution needs to be heated to assist encapsulation of the larger species.

Note that a typical synthetic procedure produces ammonium ion complexed CB[5] (CB[5]·$2NH_4$) in which an ammonium ion binds at each portal of CB[5]. The procedure to prepare free CB[5] was described in Chapter 2.

3.2.2. Cucurbit[6]uril

Cucurbit[6]uril (CB[6]) has a wider portal and a more voluminous cavity than CB[5]. CB[6] uses both its portals and its cavity to form stability inclusion complexes. Mock pioneered the host–guest chemistry of CB[6] by surveying many different complexes. Mock especially examined the complexes between CB[6] and aliphatic and aromatic amines.[9,41] Buschmann extended this work to measure equilibrium constants and thermodynamic data for CB[6] complexes with ω-amino alcohols,[171] aliphatic alcohols, acids and nitriles,[172] detergents,[173] α-amino acids and dipeptides.[174]

These studies revealed that CB[6] is particularly good at forming complexes with aliphatic ammonium species. The protonated amines bind very well to the electronegative portal, and the hydrophobic alkyl chain occupies the hydrophobic cavity, while also displacing water molecules from the cavity. There is a size-complimentary aspect to the binding affinities: alkyl diammoniums that can comfortably bind both portals and thread their alkyl chain through the cavity form the highest stability complexes (Figure 3.3). Indeed, some of the strongest CB[6] complexes known are of this type, 1,6-hexane and 1,5-pentanediammonium (2.9 and 1.5×10^8 M^{-1}, respectively, both in 50 mM NaCl).[175] The size selectivity is very high; the affinity drops dramatically when the carbon chain is shorter than four methylene units (2.0×10^7 for 1,4-butanediammonium

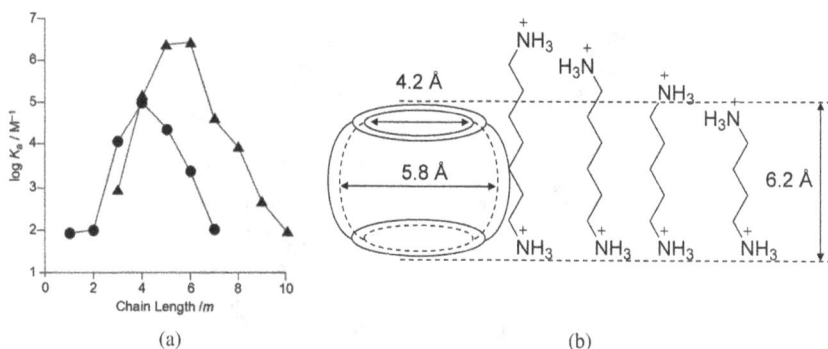

Figure 3.3. (a) Plot showing binding affinity against chain length for mono-(•) and di-ammonium (▲) alakanes. (b) The length of alkyl chains compared to the height of CB[6]. (Modified from Ref. 132)

and 3.3×10^2 for 1,3-propanediammonium, in 50 mM NaCl).[175] Since the 1,3 propane diammonium cannot reach both portals, it does not form an inclusion complex; instead, it binds to the exterior.[176] Longer alkyl diammoniums also tend to bind to one portal only to avoid entropic penalties resulting from a restrictive conformation. A representative X-ray structure is shown below (Figure 3.4).

The highest affinity guests uncovered by Mock for CB[6] are those with additional alkylammonium groups such as the biogenic amines spermine and spermidine.[41] He proposed that the CB[6] occupies the alkyl chain between the two central ammoniums and that the additional alkyl ammonium arms fold back to make extra ion–dipole interactions with portal. This binding mode was later confirmed by X-ray crystallography.[176] The resulting binding affinities increase by roughly an order of magnitude per ammonium group: Spermine > spermidine > 1,4-butanediamine. The binding affinities between CB[6] and spermine can be as high as 10^{10} M^{-1} depending on the medium in which it is measured.[175]

Alkyl (di)amines are common motifs in nature. Indeed some of those discussed above are found in nature. Acetylcholine is another biogenic amine that CB[6] is a receptor for,[134] and it is also an important neurotransmitter. As such, CB[6] has been used as a biosensor for several of these amines, including acetylcholine and several diamines (see Chapter 4.3 for sensor applications).

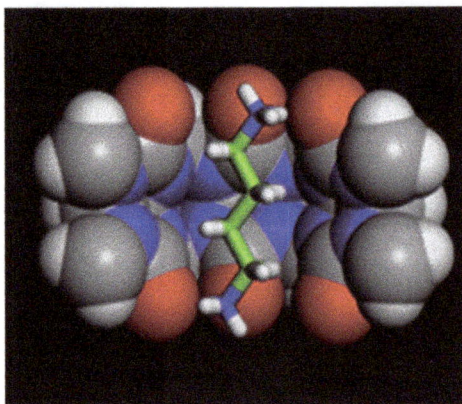

Figure 3.4.   X-ray crystal structure of 1,4-diammoniumbutane-CB[6] complex. (From Ref. 176)

Despite its small internal volume, CB[6] can incorporate aromatic guests. Five-membered aromatic rings such as thiophenes are readily encapsulated, as are six-membered ring aromatics such as benzene and xylene diammonium.[9] The crystal structure of the latter complex shows that CB[6] becomes ellipsoidal upon guest encapsulation.[177]

CB[6] has a limited ability to form ternary complexes, where it encapsulates two guests in its cavity. CB[6] can form 1:2 host–guest complexes such as 1-ethyl-3-methylimmidazolium,[178] where the guest arrange themselves in such a way that their ethyl groups point into the cavity. Catecholamines such as adrenaline[179] and isoprenaline[180] also exhibit this kind of behaviour. Diastereomeric ternary complexes can be formed with chiral guests.[61] For example, chiral cyclohexylamine can bind to the exterior of CB[6], and the complex preferentially encapsulated one enantiomer of methylbutylamine over another with 95% *ee*.

CB[6] is capable of binding a small range of neutral guests too. Gases such as sulphur hexafluoride[181] and xenon[47] can form complexes (3.1×10^4 M^{-1} and $2.1 \times 10^2\,M^{-1}$, respectively, in 0.2 M Na_2SO_4) with CB[6]. CB[6] can form complexes with a range of hydrocarbons, in 1 mM HCl, some with binding in the 10^5–$10^6\,M^{-1}$ range.[83] Some of the halogens, Br_2 and I_2, are also encapsulated in CB[6].[182]

Like CB[5], CB[6] binds to various alkali, and alkaline earth metals, including Na^+, K^+, Rb^+, Cs^+, Ca^{2+} and Sr^{2+}. The diameter of Cs^+ is an almost perfect fit for the portal, which allows Cs^+ to interact with all the carbonyls in the portal.[183] *i*CB[6] is an isomer of CB[6] with one glycoluril group pointing inwards, which reduces the volume of the cavity. It binds some of the same guests as CB[6] does, but with lower affinity and kinetic lability.[57] To illustrate, diammoniumhexane binds to *i*CB[6] with only millimolar affinity, whereas it binds to CB[6] with micromolar affinity.

3.2.3. *Cucurbit[7]uril*

Cucurbit[7]uril (CB[7]) stands out amongst the other members of the cucurbituril family because it exhibits some of the highest binding constants reported for synthetic host–guest pairs. The strongest reported guest to date is diamantane diammonium which has an attomolar dissociation constant.[33] CB[7] also forms high-stability (10^7–$10^{17}\,M^{-1}$) complexes with

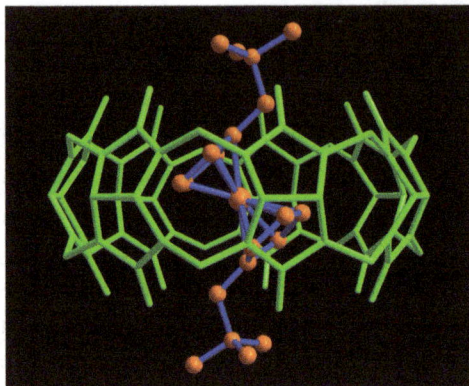

Figure 3.5.   X-ray crystal structure of a CB[7]-bis(trimethylammonium)methylferrocence. (From Ref. 32)

adamantyl-, metalloceneyl-, *p*-xyleneyl- and trimethylsilyl-containing guests.[30,31,184] A representative X-ray structure can be seen in Figure 3.5. The highest affinity guest for CB[7] is diamantane-based guest, which has an attomolar dissociation constant.[33] This guest was designed to fill as much space in the cavity as possible and optimise the geometry of the ion–dipole interactions.

Some guests exhibit remarkably high selectivity for CB[7]. For example, adamantaneammonium has a 5000-fold higher affinity for CB[7] than for CB[8]. However, the addition of two methyl groups to the adamantyl core completely flips the selectivity; the methylated analogue has a 10^7-fold preference for CB[8] over CB[7]. There was also tremendous functional group selectivity, while adamantaneammonium forms a complex ($K_a \sim 10^8$) with CB[7], changing the ammonium group to a carboxylate means that no complex is formed at all. Such selectivity in water is unprecedented outside of biochemical systems and is one of the ways that cucurbiturils stand apart from other synthetic receptor molecules.

The voluminous cavity allows CB[7] to encapsulate many different cationic guests (even those with diffuse positive charges) with good affinity (around 10^6 M^{-1}) rather than forming complexes around its portal.[185] Molecules with a diffuse positive charge recruit water molecules to solvate their charge; desolvation of the guest and cavity represent a powerful driving force of complexation. Cationic tricyclic aromatic dye molecules tend to have delocalised charge and many of these are guests of CB[7]. Furthermore, photochemical behaviour of encapsulated dye molecules may change, and this has been exploited to probe the properties of the cavity (Chapter 2.3) and in several other applications (Chapter 4.2). A variety of cationic drug molecules of appropriate size have been encapsulated in CB[7], in particular. These are discussed in Chapter 7.

The appreciable solubility of CB[7] in water and its cavity size has allowed the exploration of host–guest chemistry with biomolecules. The aromatic amino acids, in particular, phenylalanine ($K_a \sim 10^7$ M^{-1}), are good guest molecules for CB[7].[186] The aromatic amino acids have the features that are necessary for recognition. Namely, they have a positive charge (at an appropriate pH) on the amine to bind the portal and a hydrophobic group to fill the cavity. Urbach *et al.*[24,187,188] have shown that this recognition is retained in peptides and even small proteins, such as insulin. However, the stability constant is weaker when the phenylalanine is not the N-terminal residue, due to the lack of an ammonium group. Binding to amino acids, proteins and peptides is discussed in more detail in Chapter 7. Another class of biomolecule that can bind to CB[7] are the amino-saccharides.[189] Interestingly, CB[7] has a remarkable selectivity for the α-anomer over the β-anomer.

*i*CB[7] is an isomer of CB[7] with one glycoluril group pointing inwards, which reduces the volume of the cavity. It binds some of the same guests as CB[7] does but with lower affinity. For example, trimethylaminemethylferrocene binds 10^6-fold less strongly to *i*CB[7] than it does to CB[7]. *i*CB[7] has a preference for flatter (aromatic) guests rather than liner aliphatic amines.[57] For example, *p*-xylenediammonium is preferred over diammoniumhexane and even ferrocene-based guests.

3.2.4. Cucurbit[8]uril

Cucurbit[8]uril (CB[8]), like CB[7], also has some strong affinities for hydrophobic molecules with cationic groups (up to 10^{14} M^{-1}).[111] CB[8] is large enough that it can incorporate smaller macrocycles into its cavity, such as cyclen and cyclam; these macrocycles can incorporate their own guests to create "Russian Doll"-type complexes.[190] Long alkylammonium chains arrange themselves into a U-shape in the cavity. The alkyl chains arrange themselves in such a way as to displace the maximum amount of water from the cavity and make van der Waals interaction with the cavity wall[191,192]; in some cases the positive charges are at the same portal.[193] An intriguing class of molecules that are encapsulated by CB[8] are EPR-active, stable nitroxide radicals, such as amino-functionalised TEMPO.[194] Encapsulation by CB[8] results in a change to the hyperfine splitting in their electron paramagnetic resonance (EPR) spectra. The most stable binary complexes are based on isodiamantane scaffold.[111] Unlike the CB[7] guests, the steric bulk of CB[8] guest is distributed horizontally rather than vertically to best fill the volume of the cavity. The highest affinity of these guests has a binding affinity of 10^{14} M^{-1} with scope for further improvement.

The standout feature of CB[8] is its ability to form ternary complexes, with either 1:2 (host:guest) or 1:1:1 (host:guest1:guest2) stoichiometry with a wide variety of guest molecules. The first of these was a homo complex of 2,6-bis(4,5-dihydro-1*H*-imidazoyl-2-yl)naphthalene.[19] Binary complexes were not observed, which suggested that the second guest incorporation is much more favourable than the first.

Hetero complexes, in particular charge-transfer (CT) complexes, where one guest is a π-donor, and the other is a π-acceptor, form particularly stable ternary complexes. For example, methyl viologen (MV^{2+}) and 2,6-hydroxynaphthalene form a stable CT complex with CB[8] spontaneously (Figure 3.6).[26] While they are termed CB[8]-stabilised CT complexes, there is some doubt about whether the complex formation is CT-driven. Recent work suggests that solvation and electrostatic interactions are more important in the formation of such complexes.[195]

CB[8]-stabilised CT complexes have many uses, for example in the construction of supramolecular architectures and biomolecule recognition. When these CT moieties are part of the same molecule separated by a spacer, they can be used to make much larger and responsive structures; these are discussed in Chapter 5. Aromatic amino acids, such as tryptophan, can serve as π-donors and are therefore the second guest in a charge-transfer complex; this feature can be applied to recognition and modification of amino acids, peptides and proteins, which is discussed in Section 7.1.

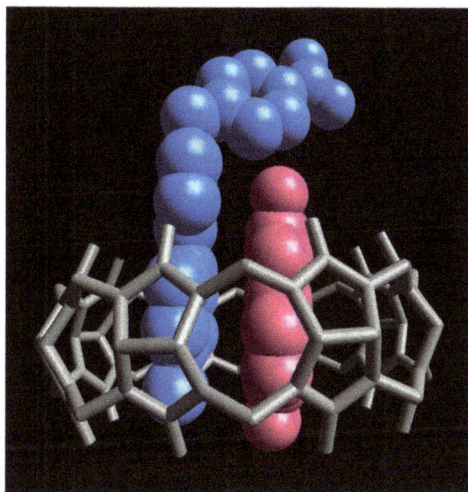

Figure 3.6.   X-ray crystal structure of CB[8] charge-transfer (CT) complex composed of a viologen derivative (blue) and a naphthalene (pink). (From Ref. 26)

3.2.5. *Cucurbit[10]uril*

Cucurbit[10]uril (CB[10]) is the most voluminous of the known cucurbituril family, and as such accommodates large guests such as other macrocycles. One example is the CB[5]@CB[10] complex, which CB[10] was first isolated as. This complex was named gyroscane since the two macrocycles rotate around each other; this complex also appears to incorporate a chloride anion inside CB[5] (Figure 3.7).[196] More interestingly, ammonium-derivatised calixarenes can also form a macrocycle within a macrocycle with CB[10]. The incorporation of the calixarene in the CB[10] alters the calxarene's host–guest chemistry, so that it binds guests that the free form would not; this can be thought of as allosteric binding.[58] Both metalated and free porphyrins bearing a methylpyridium can also be encapsulated by CB[10].[197] More recently, a viologen-based macrocycle was encapsulated in CB[10] and the viologen macrocycle then formed a charge transfer complex with hydroxynaphthalene.[198]

Other guests of CB[10] include triazene–arylene molecules.[199] Interestingly, these molecules arrange themselves so as to maximise interactions with the CB; there are no intramolecular π–π interactions in the

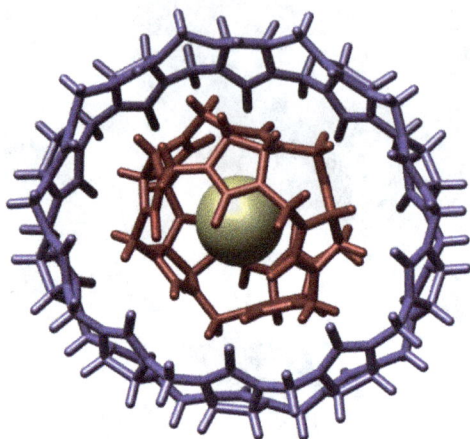

Figure 3.7.   X-ray crystal structure of CB[5]@CB[10] complex. (From Ref. 196)

triazene–arylene guests while they are in the CB. Several organometallic complexes including anticancer compounds[200] and luminescent iridium complexes[201] have also been encapsulated inside CB[10].

3.2.6. Twisted CB[13–15]

Cucurtbit[14]uril (CB[14]) adopts a twisted figure of eight-type structure and so is less voluminous than some of the smaller family members. It seems to behave like two connected smaller cavities, rather than one large cavity. Little is known about its host–guest chemistry at present, because it was only discovered recently by Tao *et al.*[99] It is known to be very flexible and can rearrange itself around a guest, to give rise to chiral complexes. For example, when it binds large lanthanide cations, like europium, it rearranges itself to form a shell-like structure.[99] The Eu^{3+} sits between the cavities, interacting with the portals, and does not sit in the cavity itself. Similar behaviour is also observed with bis-*N*-alkyl-substituted viologen guests.[127] The pyridiniums sit in between the cavities, the cationic region interacting with the portals, and the alkyl chains sit in the cavity making hydrophobic interactions. These complexes have association constants in the 10^6 M^{-1} range. Cucurbit[13]uril and Cucurbit[15]uril have recently been synthesised and only preliminary host–guest chemistry is known. Like CB[14], these new members bind heavy metal ions such as Dy^{3+} and Cd^{2+}.

3.3. Thermodynamics and Kinetics of Host–Guest Binding

In this section, the thermodynamic and kinetic considerations involved in controlling and optimising host–guest binding are examined. These include the role of water molecules, optimisation of ion–dipole interactions and the shape and size complementarity between host and guest. There are several ways to measure biding affinities, for example, NMR competition experiments with other guests. Multistep ITC experiments with increasing affinity guests is also a popular method.

3.3.1. *Thermodynamics*

Cucurbiturils are different from other host molecules because they bind many of their guest molecules with very high affinity. There are three main reasons why CBs can make such high-affinity interactions. The first is that they can offset the entropy–enthalpy compensation principle; the entropy of formation is often net favourable, whereas in other systems it is usually a penalty; CDs demonstrate this principle very well. The large deviation from enthalpy compensation is best exemplified by the ultrahigh affinity guests, but is also seen in moderate affinity binders too, such as methyl viologen (MV). Generally, this is because of the number of water molecules displaced from the portal and the cationic guest after binding. The second is that hydrophobic interactions are strongly exothermic because of high energy water molecules in the cavity; the water molecules inside the cavity experience energetic frustration because they cannot form many H-bonds in the hydrophobic cavity. Guest binding releases these water molecules and relieves their energetic frustration. The third is that rigid host molecules allow guests to be designed that can maximize size and shape selectivity; other hosts are more plastic and therefore more difficult to design optimal guests for. Some of the informative structures and corresponding thermodynamic data are given in Table 3.1.

High-stability host–guest complexes of CB[7] are informative for interrogating the thermodynamic properties of CB[n] binding. The ferrocene

Table 3.1.   Thermodynamic data for selected guests of CB[7].

Guest	K [M^{-1}]	Enthalpy, $\Delta H°$ [kcal mol^{-1}]	Entropy, T$\Delta S°$ [kcal mol^{-1}]
MV	2.0×10^5	−3.4	3.9
F1[a]	$(3.2 \pm 0.5) \times 10^9$	−21.5 ± 0.5	8.6 ± 0.5
F2[a]	$(4.1 \pm 1.0) \times 10^{12}$	−21.5 ± 0.5	4.3 ± 0.4
F3[a]	$(3.0 \pm 1.0) \times 10^{15}$	−21.5 ± 0.2	0.5 ± 0.5
S1[b]	$(1.5 \pm 0.1) \times 10^{12}$	−15.4	1.43
S2[b]	$(3.2 \pm 0.2) \times 10^{12}$	−15.4	1.87

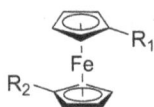   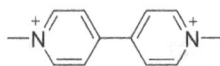   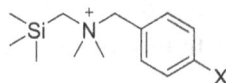

F1 R$_1$ = CH$_2$OH,   R$_2$ = H   **MV**   S1 X = H
F2 R$_1$ = CH$_2$NMe$_3$,   R$_2$ = H   S2 X = NO$_2$
F3 R$_1$ = CH$_2$NMe$_3$,   R$_2$ = CH$_2$NMe$_3$

derivatives **F1–3** are either neutral or have varying amounts of positive charge, yet their enthalpy of binding is very similar ($\Delta H° = -21$ kcal mol^{-1}).[32] Instead, the total binding entropy becomes more favourable with each positive charge added to the ferronce. To illustrate, $-T\Delta S = 8.6$ kcal mol^{-1} for neutral ferrocene, **F1**; 4.3 kcal mol^{-1} for single-charged ferrocene, **F2**; and 0.5 kcal mol^{-1} for double-charged ferrocene, **F3**. Thus, it seems that the high-stability cucurbituril host–guest complexes do not obey the enthalpy–entropy compensation effect that is prevalent in other supramolecular systems.[174]

The enthalpy–entropy compensation effect describes how molecular recognition enthalpically drives the formation of most host–guest complexes, but an entropic penalty is paid because of reduced configurational flexibility.[202] However, thermodynamic data for high-affinity CB-guests suggest that this principle is not obeyed. To unravel why this is the case, the entropy was separated into its component parts (configurational entropy and solvation entropy).[32] While a penalty is paid with configuration entropy, solvation entropy is positive which indicates that solvent molecules are returned to the bulk not only from the cavity but also from the heavily solvated portal and cationic guests. Taken together, the desolvation of the cavity, portal and guest helps the host–guest pair overcome the enthalpy–entropy compensation principle.

Another surprising observation that needs explanation is why are the ion–dipole interactions enthalpically neutral? Computational modelling suggests that there are two finely balanced processes at play: the ion–dipole interaction formed between the guest and the CB[7] portal is enthalpically favourable, but this is offset by the loss of solvation interactions between water molecules and the cationic guest and the polarised portal. The net result is that the formation of these ion–dipole interactions is driven entirely by entropy.[203] Experimental evidence has also supported that the formation of an ion–dipole interaction is enthalpically neutral.[204] Using various *N*-substituted benzyl-trimethylsilylmethylammonium guests, a Hammett study was performed. By measuring the affinity of probe molecules with X = H and X = NO_2 (**S1** and **S2**, respectively) it was found that the binding affinities to CB[7] were subtly different (2.5-fold difference), but the difference resulted from entropy, not enthalpy.

Water molecules that occupy the cavity also play an important role in driving complex formation. The observation that neutral ferrocene binds with an affinity of 10^9 M^{-1} is remarkable considering that β-cyclodextrin (a similarly sized macrocycle that also binds through hydrophobic interactions) complexes rarely exceed 10^6 M^{-1}.[205] This suggests that the hydrophobic effects in this system are enthalpically dominated.[206] Classical hydrophobic interactions are driven by increased entropy by displacing water molecules. However, displacement of water molecules from molecular containers is enthalpically favourable because the water molecules inside are restricted in the number of H-bonds they can make; barrel-shaped containers such as CBs amplify this effect more than more open containers. Molecular dynamics simulations show that water molecules inside the CBs are very restricted in the number of H-bonds they can make.[89] Water molecules cannot make many interactions with the walls of the cavity either. In the case of a narrow container, such as CB[5], there are few water molecules inside. In a larger container of CB[6] and CB[7], there are some energetically frustrated (so-called high-energy) water molecules and displacement of these water molecules by guest binding is strongly exothermic. As the container gets even larger, there is little difference between the bulk solvent and the encapsulated solvent. The experimental data reflect this trend: the exothermic driving forces are higher with CB[7] complexes than those of CB[6] or CB[8].

The ternary complexes of CB[8] also display an enthalpically dominated hydrophobic effect. The inclusion of the first guest e.g. methyl viologen restricts the remaining water molecules in the cavity, thereby increasing their energetic frustration of the remaining water molecules. Consequently, the second guest binding is strongly exothermic ($\Delta H° = -15$ kcal mol^{-1}) since those high-energy water molecules are ejected.[207] Additional evidence can be found by examining solvent and solvent isotope effects: the binding affinity of the second guest drops more than 1000-fold when the solvent is acetonitrile, showing the involvement of water molecules.[208]

A note on terminology: The term "high-energy water" is used here because that is the term prevalent in the CB[n] literature to describe energetically frustrated water molecules inside the CB[n] cavity; it also helps to understand the thermodynamic processes. However, other chemists may talk about this effect in terms of a drying transition or the dewetting potential that the cavity possesses.

3.3.2. Kinetics

For CBs to be incorporated into responsive materials or molecular machines, it is necessary to understand the factors that control the binding mechanism of the host–guest interaction. The cavity of CB[n] has a wider diameter than that of its portal, which means that the portal is a kinetic barrier to guest ingress and egress. This manifests as a slow association and even slower dissociation rate, due to the constrictive binding mode. This is in contrast to many other macrocyclic host molecules, such as calixarenes and cyclodextrins, the "portals" of which are wider than the cavity itself. Cucurbiturils can form exceptionally tight complexes because of this constrictive binding.

In the 1980s, Mock and Shih began investigating the binding mechanism with CB[6] by measuring the dissociation rates by ^{1}H NMR.[41,209] Among the guests surveyed were cyclopropyl-, cyclobutyl- and cyclopentylmethaneammonium; these three molecules revealed an interesting trend: the wider the guest, the slower the complex formation. This observation makes sense when the narrow portal diameter, compared to the inner cavity, is considered. A UV-vis spectroscopy study on CB[6] and benzene-based guests also suggested the existence of two different complexes, namely an exclusion and inclusion complex.[43] Several years later, Nau *et al.*[53,210] investigated the binding mechanism in detail. They used cyclohexylmethylammonium as a probe molecule since it forms complexes on the order of hours to days, which allows for detailed spectroscopic measurements of the process. They performed pH-dependent measurements of $k_{ingress}$ and found that the rate ingress is higher when the guest is unprotonated, which, they suggested, pointed to the involvement of free amines in the binding mechanism. They proposed that there are two pathways by which a protonated host–guest complex can form. The free amine guest can enter the CB[6] cavity and then be protonated; or, at a pH below the pK_a of the guest, the alkyl ammonium forms an exclusion complex with the portal with the alkyl chain still in the bulk solution. Then a "flip-flop" mechanism occurs allowing the alkyl chain to move into the cavity (Figure 3.8). They also discovered that the rates of ingression were slower in the presence of metal or other cations.

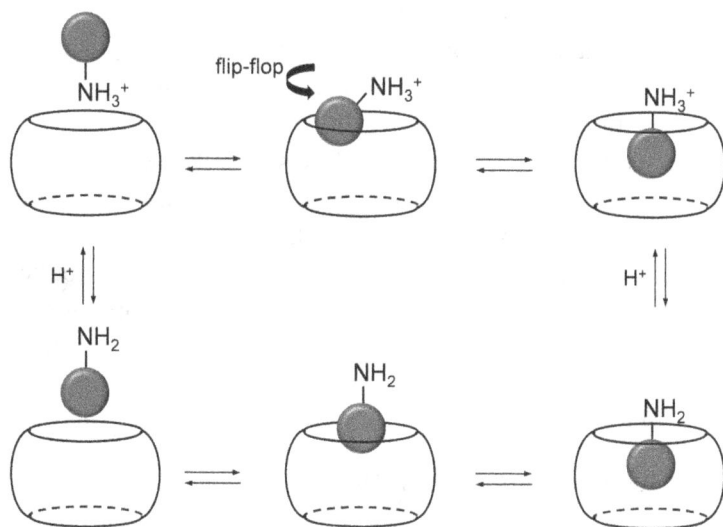

Figure 3.8. Binding mechanism of sterically demanding alkyl ammoniums to CB[6]. (Modified from Ref. 53)

Kinetic measurements of some CB[6] complexes reveal that it can form exceptionally tight complexes. For example, the diammoniumcyclohexane@CB[6] complex has a k_{on} and k_{off} of 1.2×10^{-3} M^{-1} and 8.5×10^{-10} M^{-1}, respectively.[66] These values correspond to a half-life of around 26 years — 100-fold slower than biological benchmark binding pair: biotin–avidin.

CB[7] seems to form host–guest complexes in a single step. Stopped flow measurements on the binding between a naphthyl guest and CB[7] revealed that the rate of association is an order of magnitude slower than a diffusion-controlled process, but much faster than the CB[6] process mentioned earlier, which is consistent with the CB[7] portal being more flexible than that of CB[6].[211] There was no evidence of any exclusion complex formation. The dissociation rates of the CB[7]-naphthyl complex were compared to that of β-cyclodextrin-naphthyl complex and it was found that the complex dissociation rate was much slower for CB[7], which indicated constrictive binding.[211] Other studies have shown that CB[7]-complex formation occurs in a single step, for example CB[7]-methyl viologen[212] and CB[7]-berberine.[213]

There are still many unanswered questions related to the binding mechanism of many CB[n] complexes. It may be the case that different mechanisms occur depending on the characteristics of the guest. Straight chain dialkylammoniums seem to enter CB[6] without hindrance, whereas a bulky guest such as cyclohexylmethylammonium enters slowly in a two-step process. In the case of CB[7], only single-step processes have been observed. How sterically demanding guests, such as ferrocene form a complex is unclear. They may be able to form cationic exclusion complexes with CB[7], but as Masson points in his review,[7] their core is rigid so the flip-flop mechanism may not be plausible. At the moment, there have been no published reports of the binding kinetics involving CB[7] and high-affinity guests such as ferrocene and adamantane based guests.

3.4. Host–Guest Chemistry in the Gas Phase

In the previous sections, the host–guest chemistry of CB[n] in solution was described, which highlighted the importance of solvation in these processes. In the gas phase, there is no solvent, so the ion–dipole inter-actions and other intrinsic guest properties dominate the host–guest chemistry.[214] Mass spectrometry is the primary tool used to investigate supramolecular systems in the gas phase. The simplicity of the gas phase is advantageous because it makes possible the comparison of theoretical and experimental work directly and it can help unravel how subtle differences in structure influence host–guest chemistry.[215]

Dearden *et al.*[215] led the early investigations into the gas-phase host–guest chemistry of CB[*n*]. A key challenge is determining whether a complex is an inclusion complex or an exclusion complex. In solution, this can be determined spectroscopically; in the gas phase, it is not so easy. One way is to perform a guest exchange with *t*-butylamine (which does not penetrate the cavity), an exclusion complex will exchange quite readily, but the exchange is slower with an inclusion complex.[216] Inclusion complexes also require higher energy to dissociate them inside the MS instrument.[50,216] Using these techniques, it has been shown that CB[5] can form an exclusion complex with 1,4-butanediammonium, where only one positive charge on the guests interacts with the CB[5] portal. However, with CB[6] the same guest forms a doubly charged inclusion complex.[50] Interestingly, there is a size selectivity for the length of the diammonium, but it is not the same as in solution. In the gas phase, the ideal length for the diammonium is four carbon atoms, whereas in solution it is six. The reason for this is likely because the six-atom linker enables more desolvation of the CB cavity and portal. In the solventless gas phase, there is no such consideration, and the best guest is the one that can optimize the ion–dipole interactions[217] Similarly, lysine forms a singly charged exclusion complex with CB[5] and forms a doubly charged pseudorotaxane-like structure with CB[6].[218] Additionally, CB[6] is a useful tool for recognising lysine in peptides since the lysine–CB[6] complex can be observed at low collision activa-tion energies; at higher energies a distinctive fragmentation pattern is observed.[219]

The small decamethylCB[5] forms cage complexes with neutral guests and ammonium species. Neutral guest molecules such as O_2, N_2, acetonitrile or methanol form inclusion complexes with $Me_{10}CB[5]$ and the ammonium molecules act as lids at both portals to entrap the guest molecule. Once the ammonium is removed from the system, the guest quickly leaves.[48] In a similar study, CB[7] can encapsulate larger neutral guests (such as benzene, fluorobenzene and toluene) which are then encased with cationic guanidinium units on either portal.

The principal strength of examining host–guest chemistry in the gas phase is that solvation effects can be excluded which leaves the intrinsic properties of the host and guest to dominate the recognition; indeed, gas-phase chemistry has provided useful insights into CB[n] complex formation. Several examples are presented below. The thermodynamic binding properties of CB[7] with amino acids was compared in the aqueous and gas phases.[186] It was found that in the solution phase, the binding was strongest with phenylalanine, whereas in the gas phase, the binding was strongest with histidine (Figure 3.9). These results emphasise the importance of solvation effects in solution and the dominance of ion–dipole interactions in the gas phase. The differences in binding between five neutral isomeric saccharides was studied in the gas phase.[220] These molecules were weak binders in solution, which made them difficult to study. In the gas phase, complexation was promoted, and in the absence of solvation effects, the subtle differences in the saccharide structures were manifested in the strength of the host–guest complexes. Gas-phase studies on complexes between CB[n] and various alkyldiammoniums revealed that water plays a key role in the high-affinity binding of these complexes.[214] In the solventless gas phase, making ion–dipole interactions sometimes causes the alkyl chain to adopt awkward conformations, whereas in the aqueous solution water molecules can make H-bonds with the ammoniums to help stabilise the complex. These water molecules play a key role in the high stability of such complexes.

Reactivity inside CBs in the gas phase can be predicted on the basis of packing coefficients (Figure 3.10).[221] In the molecules shown below, cycloelimination reaction was favoured when the packing coefficient was 30–50%, i.e. when the guest was constricted by the CB, but just short of the ideal 55%.[222] Furthermore, there must be sufficient void volume for the reaction to take place since the cycloelimination is a positive volume reaction. When the binding was too loose, the complex dissociated.

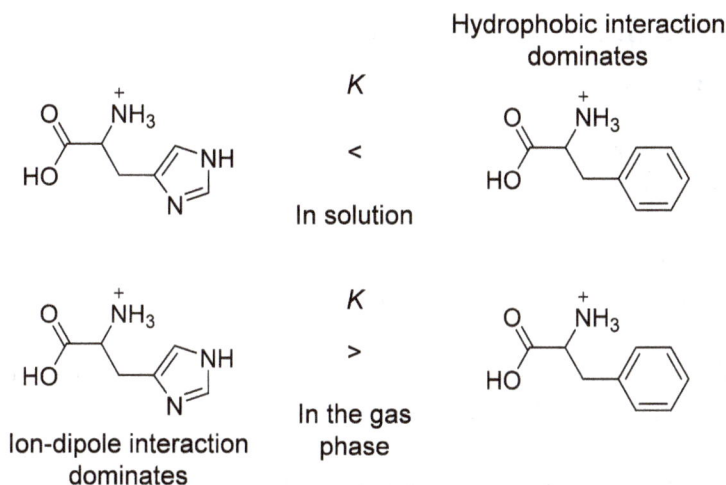

Figure 3.9. Different preferences of CB[7] for amino acids in the gas phase and in solution.

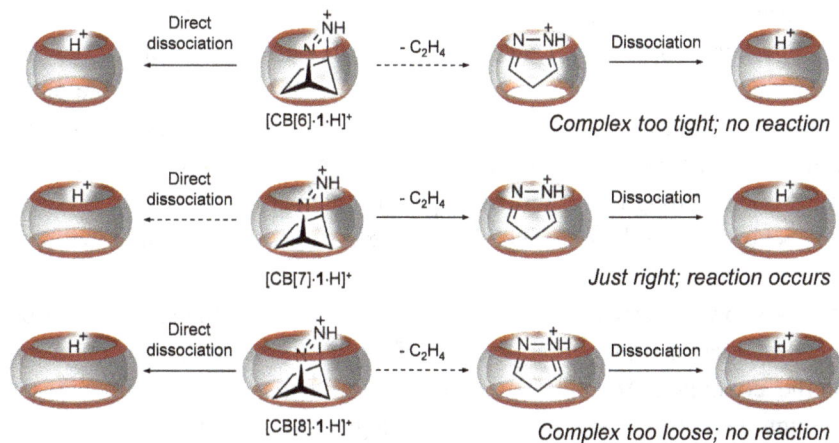

Figure 3.10. The gas–phase reactivity of CB[n] with a small guest. (Modified from Ref. 221)

When these conditions were not met, the reactivity within the cavity was significantly reduced. These results have important implications for reactivity not only for cucurbiturils but also for chemistry in confined spaces more generally.

3.5. Host–Guest Chemistry of Cucurbituril Analogues

3.5.1. *Acyclic cucurbiturils*

The Isaacs groups have pioneered the development of acyclic analogues of cucurbiturils. After several generations of compounds, they found that oligomeric glycolurils capped at both ends with an aromatic group can act as high-affinity hosts for molecules that are guests for CB[6] or CB[7].[149] The cavity volume of these hosts is similar to that of CB[7]; they bind guests of CB[6] and CB[7] with affinities in the range of 10^5–10^9 M^{-1}. Like CB[n], increasing the number of ammonium moieties increases the binding affinity. Unlike CB[n], there is only a moderate dependence on the length of alkyl chain and the alkylation state of the ammonium. This moderate selectivity is most likely a result of the flexibility of the host and the conformational changes required to make the complex.

The flexibility and the synthetic tractability of the host allows extensive modulation of the host–guest chemistry and physiochemical properties. The aromatic groups can be functionalised by adding sulphonate groups to solubilise the containers. Most notably, sulphonate-functionalised acyclic cucurbiturils have been shown to bind and solubilise poorly soluble pharmaceuticals.[150] Many drugs are poorly soluble because they often have aromatic cores; the aromatic capping groups of the acyclic CBs can bind these drugs through π–π and hydrophobic interaction. Acyclic CBs have also been shown to bind and solubilise single-walled carbon nanotubes also by a combination of π–π interactions and hydrophobic effects.[90] The containers show selectivity for low diameter nanotubes, although this can be tuned by increasing the ionic strength of the solution and the concentration of the container. Acyclic CBs can bind unfunctionalised hydrocarbons, through the hydrophobic interactions, in aqueous solution.[107]

3.5.2. *Hemicucurbiturils*

Hemicucurbit[6]uril (*h*CB[6]) is exactly half of CB[6] (refer back to Figure 2.6). The carbonyls arrange themselves either up or down in an alternating manner, resulting in an electropositive interior. As a result, *h*CB[6] prefers to bind anions such as thiocyanate, although it has also been shown to weakly bind some cations, such as Co^{2+}, Ni^{2+} and UO$_2^{2+}$.[223,224] The very poor solubility of *h*CB[6] makes measurements of stability constants challenging.

A chiral version of *h*CB[6], (all *R*, or all *S*) cyclohexylhemicucurbit[6] uril, was synthesised by Aav *et al.*[152] They have a similar structure to their parent *h*CB[6] with their cyclohexyl groups pointing up or down in an alternating manner. These enantiomers have two regions: a hydrophobic exterior where the cyclohexyl groups are pointing outwards and a polar cavity due to the presence of the carbonyl groups in the middle. These enantiomeric *h*CB[6] bind selectively enantiomers of methoxyphenylacetic acids as well as bind simple anions and cations, albeit quite weakly compared to the analogous CB[6]. The easy access to chiral hosts may be the key strength of hemicucurbiturils compared to native CBs or other analogues.

3.5.3. *Nor-seco-cucurbiturils*

Nor-seco-CB[6] (*ns*-CB[6]) is similar to CB[6] except that there is one missing methylene bridge, which leads to a slightly distorted structure.[151] The gap left by the missing methylene group leaves some space that allows the *ns*-CB[6] derivative to accommodate larger guests than CB[6]. For example, meta-substituted methyl aromatic amines have the steric and electrostatic features to take advantage of the distorted structure. The protonated amine binds to the undistorted portal, and the methyl groups can fit into the gap in the cavity wall.

Bis ± *ns*-CB[6] is an analogue of CB[6] that is missing two methylene bridges; the structure is held together by diagonal bridges at those positions. It displays slightly different binding selectivity compared to that of CB[6] and, since it is chiral, it can form diastereomeric complexes.[68] Bis ± *ns*-CB[6] is slightly bigger than CB[6] but smaller than CB[7]. Its portal is narrower than CB[6]. Consequently, it shows higher binding affinity for flatter guests, whereas, for alkylammoniums, the binding is around 3000-fold less ($\sim 10^5$ M^{-1}). The difference in binding is probably a result of decreased ion–dipole interactions with the narrow cavity or a reduction in desolvation compared to CB[6]. With racemic mixtures of aromatic chiral amines, diastereoselective complex formation was observed (up to 88:12). With aliphatic amines, however, the guest exchange was too fast to measure by ¹H NMR.

Nor-seco-cucurbit[10]uril (*ns*-CB[10]) can be thought of as two connected cavities, each with a volume between that of CB[6] and CB[7]; consequently, *ns*-CB[10] binds guests that both these CBs can bind.[62] These cavities display allosteric binding behaviour: the binding in one cavity influences the reactivity of the other, and computational studies suggest that the cavity contracts or expands depending on the size of the guest. The orientation of the guest binding is also important because this gives rise to diastereomeric complexes. In the case of aminoadamantane, there is a clear preference for the orientation of the guest in the complex, the cation binds to the looser portal where the methylene bridge is missing in both cavities. Binary complexes are not observed, which suggests that the guest binding is cooperative, i.e. the second guest binding is much more favourable after the first is present.

3.5.4. *Bambusurils*

In contrast to cucurbiturils which prefer to bind cations, bambusurils prefer to bind anions. The first of these to be synthesised was bambus[6]uril (BU[6]).[80] Like CB[n]s, BU[n] have poor water solubility. However, they dissolve appreciably in methanol and chloroform. The molecular recognition is modulated by 12-weak C–H•X$^-$ interactions. BU[6] has yet to be synthesised anion-free, but a liberated BU[6] can be afforded by oxidation of the anion guest.[225] BU[6] binds anionic halogens in the order of I$^-$ > Br$^-$ >Cl$^-$ >F$^-$; the most weakly solvated anion binds the best, which also reflects that the larger guest fills more space in the cavity.[80] BU[6] also forms moderate stability (~10^5 M^{-1}) complexes with BF$_4^-$, CN$^-$ and NO$_3^-$ anions.[225] A representative X-ray crystal structure is shown in Figure 3.11.

The bambusurils are a family of molecules which come in a range of sizes and peripheral decorations, which affects their host–guest chemistry.[154] Bn$_8$BU[4] is a macrocycle with a small cavity which does not display any anion binding. On the other hand, Bn$_{12}$BU[6] binds I$^-$ with an affinity 3.8×10^9 M^{-1}, which is a 12-fold affinity over Br$^-$; 1200 over Cl$^-$; and 4400 over F$^-$. Replacement of the heteroatoms of the carbonyl with sulphur has been shown to increase binding towards anions.[226]

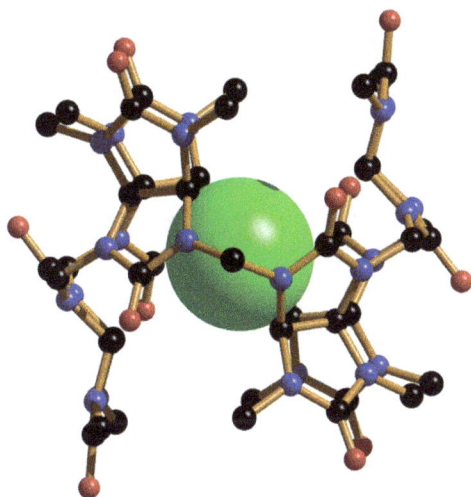

Figure 3.11.   X-ray crystal structure of a bambusuril–chloride complex. (From Ref. 80)

Water-soluble analogues, with appended carboxylates, have been synthe-sised and their binding affinities in water towards various anions have been investigated.[155] As well as binding the halogens, they also bind the toxic molecule perchlorate (ClO_4^-) with a high affinity (5.5×10^7 M^{-1}), which is exceptionally high for a neutral host in water.

Chapter 4

Cucurbit[*n*]urils as Molecular Containers and Their Applications

4.1. Chemistry Facilitated or Inhibited by CB[*n*]

CB[*n*]'s ability to encapsulate and restrict guest molecules within their hydrophobic cavity can facilitate, template, catalyse or inhibit chemical reactions. The combination of a hydrophobic interior and hydrophilic portals restricts the guest molecule's freedom of movement, in a way that favours not only reactivity but also stereoselectivity. Furthermore, CB[*n*] are useful, stabilising otherwise unstable species, and serve as protective groups because of their ability to encapsulate specific motifs.

4.1.1. *Pericyclic reactions inside CB[n]*

In the 1980s, Mock realised that CB[6] can accelerate the rate of a 1,3-dipolar cycloaddition reaction between a propargyl ammonium and an ammonium azide (Figure 4.1).[40] The copper-catalysed version of this reaction has become the archetypal click reaction.[227] Although CB[6] facilitated a significant rate enhancement (6×10^4), with regioselectivity, the reaction cannot be considered catalytic since the newly formed triazole cannot leave the cavity. This work seemed to be an example of Pauling's principle, which states that the transition state of the reaction binds to the container with greater affinity than either substrate (or product).

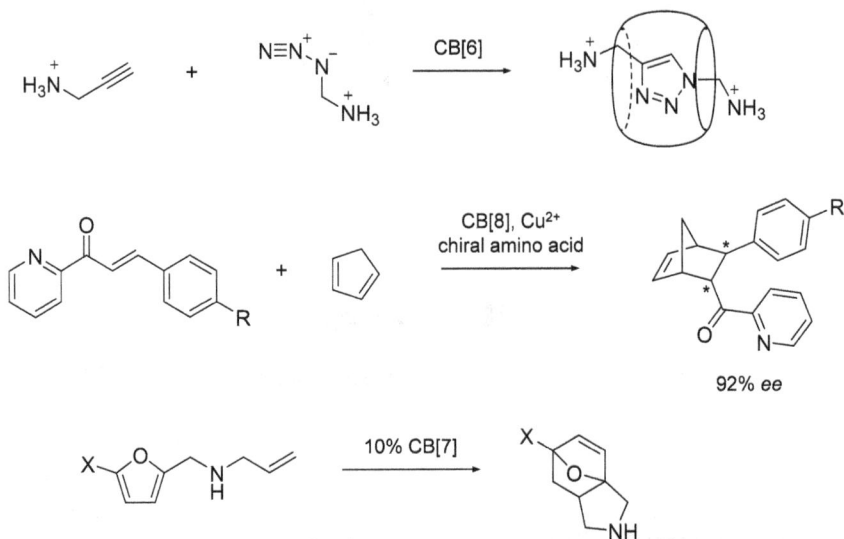

Figure 4.1.   Pericyclic reactions promoted by CB[n].

However, computational studies do not support this interpretation.[228] Instead, the rate acceleration was attributed to reduced entropic constraints through preorganisation of the substrates directed by interactions with the macrocycle, and the strain-induced compression of the reactants within the macrocycle. This work paved the way for the preparation of supramolecular assemblies based on this chemistry, such as polyrotaxanes,[229] which is discussed in the next chapter.

Diels–Alder reactions can occur in the cavity of CB[7] and CB[8]. In the case of CB[8], asymmetric catalysis was demonstrated by forming a ternary complex consisting of CB[8], a chiral amino acid and a dienophile. Cu^{2+} was added as a Lewis acid and cyclopentadiene was used as the diene (Figure 4.1).[106] The reaction occurred with up to 92% *ee*; although the reactions occurred without CB[8], there was a significant (9.5-fold) rate acceleration. A recent example demonstrated that an intramolecular Diels–Alder reaction can occur inside CB[7] (Figure 4.1).[112] Most importantly, the reaction was catalytic and occurred with as little as 10% CB[7]. Since the products bound only slightly more strongly than the substrate, there was competition between the product and the substrate allowing product inhibition to be overcome.

4.1.2. *Photochemical reactions facilitated by CB[n]*

The cavity of a molecular container is an ideal vessel for photochemical reactions. In a solvated environment, it can be difficult for the correct molecular orientation of the substrates to be achieved and the excited states can non-radiatively decay to the ground state by collisions with solvent molecules, thereby reducing reaction efficiency. CB[n] molecules can give regio- and stereo-control to photochemical reactions by placing the substrates in a confined space while also protecting them from solvent molecules. Furthermore, CB[n] are transparent to the near UV-vis region of the spectrum, meaning that photochemical processes can be promoted within the CB cavity, without interference from the container. CB[7] and CB[8] have been used in this area since they have a sufficiently large volume to accommodate two guests.

Kim *et al.*[230] were the first to investigate photochemical reactions inside CB[n]. They demonstrated that *E*-diaminostilbene can form a 1:2 host–guest complex with CB[8] in an aqueous environment (Figure 4.2). Upon irradiation with UV light (300 nm), a [2 + 2] cycloaddition occurs.

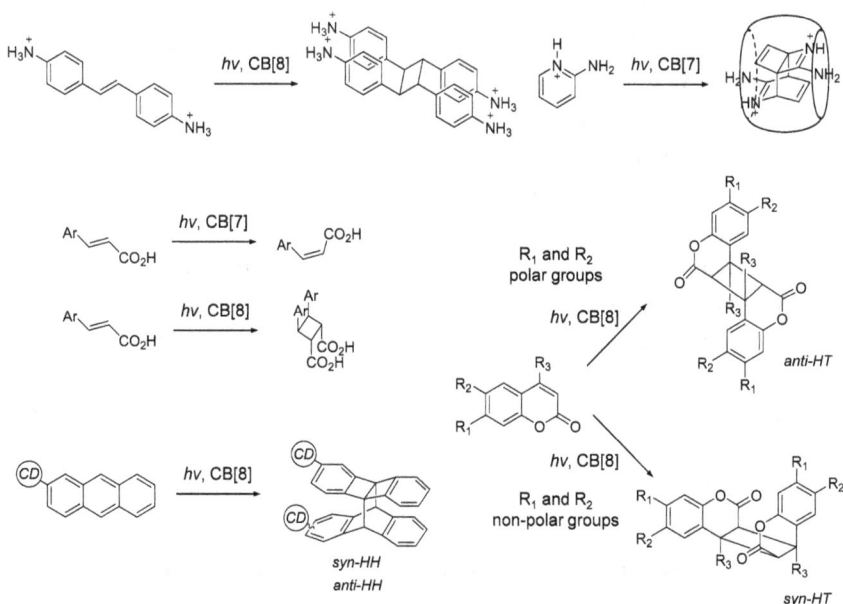

Figure 4.2. Photochemical reactions promoted by CB[n].

They observed a high selectivity for the *syn*-adduct over the *anti*-adduct. In the absence of CB[8], a photoisomerisation (*E* to *Z*) reaction occurred; this isomerisation did not occur when CB[8] was present. Furthermore, the addition of base caused the ejection of the adduct, thereby regenerating the CB[8] catalyst.

A [4 + 4] photodimerisation of aminopyridine can be mediated by CB[7], where the *anti-trans* dimer is formed exclusively (Figure 4.2).[231] Without CB[7], *anti* and *syn* products were formed in a 4:1 ratio, respectively. The origin of the selectivity seemed to be the stabilisation of the photoproduct by CB[7], which prevented the retrocyclisation back to the starting materials.

There have been several other examples of [2 + 2] photocycloaddition reactions inside CB[8].[232–235] By irradiation of stilbazole complexes in CB[8], again the *syn* dimers were obtained. Without CB[8], no photo addition reaction occurs; instead, other reaction pathways are favoured such as isomerisation or hydrolysis.[232,234] This chemistry was extended to use neutral guests, namely *E*-cinnamic acid.[233,235] A ternary complex consisting of CB[8] and two molecules of *E*-cinnamic acid was irradiated with UV light to yield the *syn*-photoadduct (Figure 4.2). However, in the presence of CB[7], only an *E*-to-*Z* photoisomerisation occurred. These examples show how CB[8] can preorganise the substrates to lead to the formation of a single isomer and that the confinement of the substrates can suppress various undesired reaction pathways.

The versatility of CB[8]-templated reactions has been increased by exploring photochemical reactions with both cationic and neutral coumarin derivatives (Figure 4.2).[236,237] A homoternary complex was formed with various coumarin derivatives, and upon UV irradiation a "head-to-tail (HT)" complex was formed. The substituents on the coumarin dictated the stereo-chemistry of the adduct formed. Coumarins with polar substituents formed the *anti*-adducts, whereas those with non-polar groups formed the *syn*-adducts. The putative reason for this selectivity is that polar molecules interact with the portal more, whereas the non-polar molecules are further engulfed in the cavity.

Other classes of photocycloaddition have been investigated, such as the [4 + 4] photocycloaddition of anthracene derivatives inside CB[8]

(Figure 4.2).[238] With anthracene-2-carboxylate, there was no preference for head-to-head (HH) or head-to-tail (HT) products. However, if the anthracene carboxylate was conjugated to α-cyclodextrin, then selectivity was observed since the α-cyclodextrin was acting as an auxiliary. They compared the selectivity of photocycloaddition reactions mediated by γ-cyclodextrins and CB[8]. In the γ-cyclodextrin case, the anticipated HT product was formed with high selectivity (98:2). However, when CB[8] mediated the reaction, not only was selectivity observed, but the selectivity was completely reversed (1:99) to the counterintuitive HH product. The proposed reason for the selectivity was that the narrowness of the cavity of CB[8] (compared to γ-cyclodextrin which is a conical frustum) prevented the HT transition state geometry being reached because the substrates cannot penetrate deeply enough into the cavity with a tethered α-CD. Whereas, in the case of γ-CD, the wider portal allows for deeper penetration of the substrates. This work demonstrated that interactions outside the cavity can also be an important factor in determining the course of the reaction.

In summary, a variety of factors affect the selectivity of the photochemical reactions that are mediated by CB[n]. The CBs serve as UV-transparent containers that protect the excited-state substrates from the solvent environment while also playing a key role in directing the selectivity of the reaction. The field of supramolecular photochemistry has been reviewed recently by Sivaguru and Ramamurthy[239]; their review discusses cucurbiturils and other classes of host molecules that facilitate photochemical reactions.

4.1.3. *Solvolysis reactions catalysed by CB[n]*

The portal of CB[n] can play a key role in regulating catalysis inside the cavity. The rate of hydrolysis of various benzoyl chlorides were compared when they were encapsulated in either CB[7] or methylated β-cyclodextrin.[240] It was found that benzoyl chlorides with electron-donating substituents were more susceptible to hydrolysis when encapsulated in CB[7], whereas they were protected by methylated β-cyclodextrin (Figure 4.3). The trend was reversed when the substituent was electron withdrawing. To illustrate, CB[7] promoted the hydrolysis

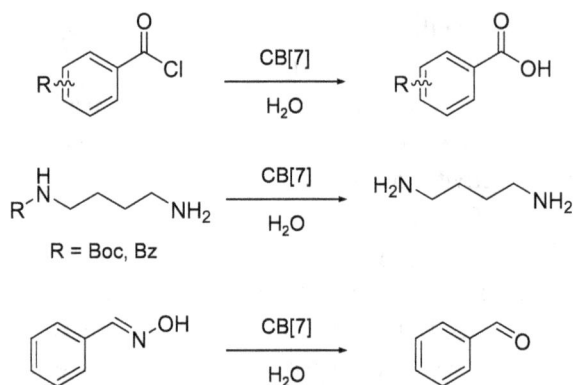

Figure 4.3.   Solvolysis reactions promoted by CB[7].

of 4-methoxybenzoyl chloride 5-fold, whereas it inhibited the hydrolysis of 4-nitrobenzoyl chloride 100-fold. The difference in reactivity can be explained by the differential stabilisation of the acylium ion transition state: the polar CB[7] portal stabilises the acylium transition state through ion–dipole interactions, but methylated β-cyclodextrin cannot offer such stabilisation.

The CB[6]- and CB[7]-mediated hydrolysis of amides, carbamates and oximes in acidic solution have also been investigated (Figure 4.3).[241] In the presence of either CB, a catalytic effect was observed. The hydrolysis of cadaverine–carbamate (which binds to both CB[6] and CB[7]) was accelerated 5-fold with both CB[6] and CB[7] at pD 0.9 (deuterated media) and 11.6-fold with CB[7] at pD 1.4. In the case of mono-*N*-Boc-cadaverine, the hydrolysis rate was 30-fold faster with CB[6] than without CB[6]. The hydrolysis of benzaldoxime was catalysed by CB[6], at pD 4, 50-fold. At a higher pD (5.8), a larger rate acceleration (285-fold) was observed. This catalytic effect was attributed to the pK_a shift (discussed in Chapter 3) upon complexation, which assists the protonation of the substrate; furthermore, the portal helps stabilise the protonated nitrogen species. While the rate acceleration was impressive, catalytic turnover was slow since the product binds 6-fold more strongly than the starting material, so product inhibition was observed.

4.1.4. *Metal cation-assisted CB[n] catalysis*

The electronegative carbonyl-rimmed portal is known to bind metal ions very well. These CB–metal complexes are beginning to be explored as active catalysts. For example, simple unbranched alkanes can be partially oxidised in the presence of CB[6]–oxovanadium(IV) complex.[242] The formation of 2-pentanol, 2-pentanone and 3-pentanone was observed, albeit with low conversion (Figure 4.4).

A novel disilylation of an alkyne reaction that is mediated by a Ag$^+$–CB[7] complex has been reported (Figure 4.4).[243] The mechanism of this reaction was postulated to be the formation of a ternary complex between CB[7], Ag$^+$ and the silyl substrate. This complex then facilitates the formation of an alkenyl–Ag$^+$ organometallic complex, which is then hydrolysed to the alkyne.

The photolysis of bicyclic azoalkanes promoted by transition metal ions coordinated to the CB[7] rim have been investigated (Figure 4.4).[244] Several CB[7]–M^{n+} complexes were tested for their photo activity with diazabicyclo[2.2.1]oct-2-ene and dizazbicyclo[2.2.1]hept-2-ene in an aqueous–pentane solvent system. Each starting material can be

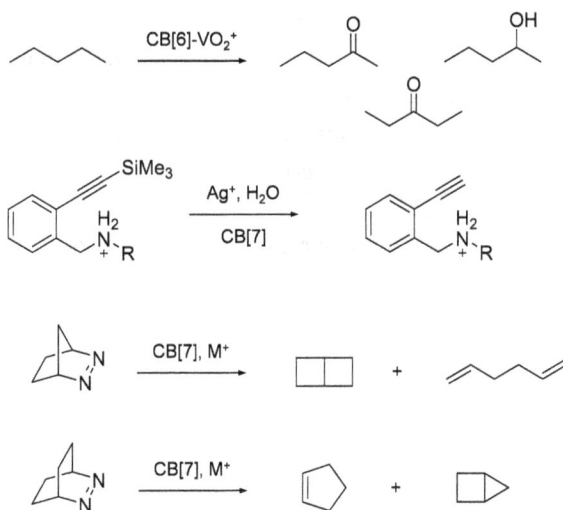

Figure 4.4.   Reactions promoted by CB[*n*]–metal complexes.

photolysed into two possible products, the ratio of which can be altered by the addition of CB[7] and a transition metal. The products of the photoreaction had a lower affinity for CB[7] and a higher solubility in the organic phase than the starting materials, which allows the catalyst to be regenerated. One interesting example is the case of Ag[+], which permits the formation of otherwise inaccessible products. The chemoselectivity was postulated to occur from the triplet excited state, populated by heavy-atom-induced intersystem crossing. This work demonstrated that supramolecular catalysis can give a route to previously inaccessible products.

4.1.5. *Stabilisation of otherwise unstable species by CB[n]*

One of the earliest and most utilised examples of CBs stabilising unstable species is the CB[8] stabilisation of a methyl viologen dimer. Single-electron reduction of methyl viologen results in formation of a radical cation which can form a weak dimer in solution. In the presence of CB[8], this dimerisation occurs in the cavity with a 10^5-fold rate enhancement and the equilibrium strongly favours the dimer (Figure 4.5).[245] This ternary complex has been used as the basis for constructing complex assemblies and molecular switches (which is discussed in Chapter 5). CB[8]-stabilised tetrathiafulvalene (TTF) dimers behave in a similar manner except that the cationic dimer is yielded by oxidation.[246] The TTF dimer in CB[8] is stable for months in aqueous solution. Both the methyl viologen and TTF chemistries are compatible with each other and can be used in the same system.[247]

CB[7] can also stabilise other radical species. An oligoaniline-CB[7] rotaxane was prepared and it was found that encapsulation dramatically

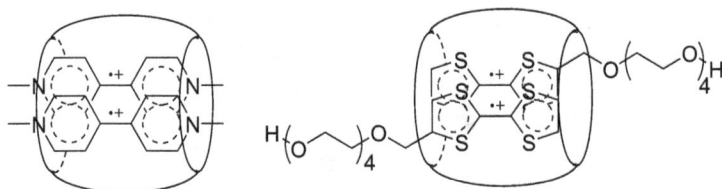

Figure 4.5.   Stabilisation of methyl viologen and tetrathiafulvalene radical dimers in CB[8].

changed the oxidation potential of the π-system and stabilised the resultant radical species.[248] The oligoaniline itself is oxidised in a stepwise manner, first to a short-lived radical species and then to a benzoquinoid. When the oligoanaline is encapsulated in CB[7], the radical species is persistent and protected from the solvent environment, and very little further oxidation was observed. The stabilisation of such a species was the basis of the preparation of a conductive CB[7]–polyaniline polymer.[78]

CB[7] can be used to slow an isomerisation reaction. *Cis*-stilbenes slowly undergo thermal isomerisation to *trans*-stilbene; however, when *cis*-stilbene is encapsulated in CB[7] the isomerisation reaction is inhibited.[249] The stabilisation is a result of the amino groups at both termini of the *cis*-stilbene forming H-bonds with both portals of CB[7]; the *trans*-isomer can only bind to one portal.

4.1.6. *Reactions inhibited by CB[n]*

Cucurbiturils ability to encapsulate guests inside the cavity can be applied in the same way as a protective group in organic synthesis. CB[n] preserve certain functional groups of the molecule through supramolecular interactions. These supramolecular protective groups offer an opportunity for more efficient protective group strategies, which are yet to be utilised extensively in organic or biochemical systems.

Disulphides that are encapsulated in CB[6] are protected from reducing and oxidising conditions. Cysteamine disulphide and derivatives of these compounds are guest molecules of CB[6]; Kaifer *et al.*[250] have shown that when CB[6] forms a complex with these molecules, the CB[6] occupies the position over the disulphide. The encapsulated disulphides were resistant to reduction by dithiothreitol; similarly, they were resistant to oxidation with $FeCl_3$, dissolved oxygen or chloropicrin (Figure 4.6). This work demonstrated the use of CB[6] as a supramolecular protective group that worked by sterically protecting the guest molecule.

CB[7] can also be used to inhibit water oxidation catalysed by a cationic Cp–Ir complex.[251] The cationic iridium complex can catalyse the splitting of water into oxygen and hydrogen. When CB[7] is added to the reaction mixture, there is a significant amount of inhibition; the mechanism of inhibition appeared to be the cationic iridium catalyst binding to the portal.

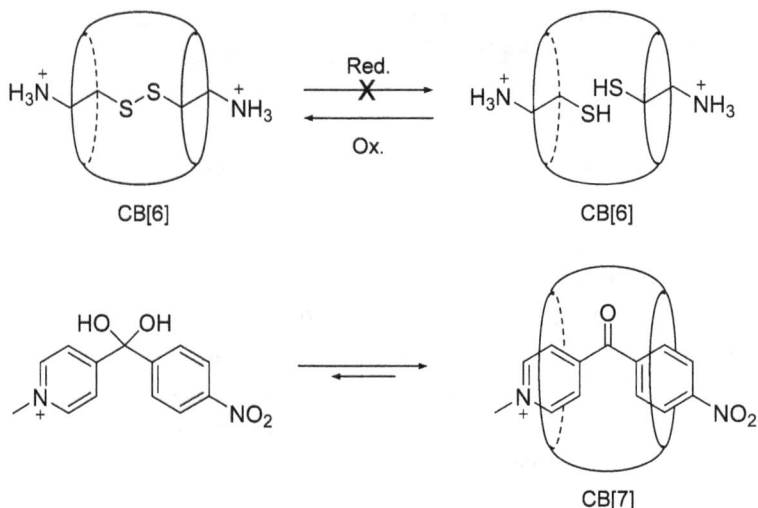

Figure 4.6.   Reactions inhibited by CB[n].

The effect of CB[7] on the equilibrium between pyridinium ketones and its hydrated *gem*-diol form has been investigated (Figure 4.6).[252] The pyridinium moiety of the guest binds to the portal and hydrophobic interactions directs the aromatic ketone into the cavity. As a result, the ketone is protected from water molecules and the equilibrium moves towards the ketone.

4.2. Dye Encapsulation and Stabilisation

The inclusion of chromophoric or fluorescent dyes into a supramolecular host can have profound effects on the dye's photophysical properties. There are several mechanisms by which macrocycle encapsulation can change the photophysical properties of a dye molecule. The most pronounced of these exhibited by cucurbiturils is the extremely non-polarisable environment of the CB[n] cavity that causes the dye to undergo a solvochromatic shift.[46] The other major difference is that encapsulated dyes are more likely to undergo *radiative* decay to the ground state, thereby increasing their quantum yield for a variety of reasons (Figure 4.7), such as: (i) they are mechanically protected from solvent molecules and chemical degradation pathways; (ii) they are conformationally restricted so they cannot easily

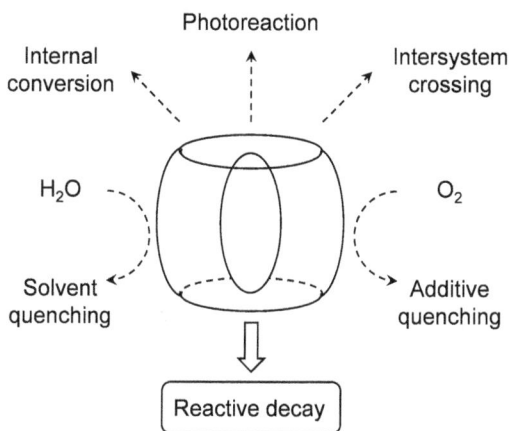

Figure 4.7.   Origin of fluorescence enhancement when dyes are encapsulated in CB[*n*]. The dashed arrows represent the processes that are often suppressed during dye encapsulation.

undergo internal conversion through rotational or vibrational motions; (iii) encapsulation suppresses dye aggregation and the subsequent self-quenching that occurs; and (iv) some dye molecules undergo pK_a shifts when encapsulated in a CB molecule.[164,165,253] Nau has reviewed fluorescent dyes and their supramolecular interactions with various hosts, including cucurbiturils.[254] (The reader is directed there for a more complete look at the field.)

Through the mechanisms described above, the fluorescence properties of dyes can be modulated by encapsulation by CB[*n*]. Enhancement of fluorescence in CB[7] is particularly strong because of the extremely low polarisability of the CB[7] cavity, compared to other CB[*n*], since the rate of radiative decay increases with the square of the refractive index of the environment.[255] The polarisability (*P*) is related to the refractive index (*n*) by Equation (4.1). The result of this is that the fluorescence lifetime can be extended; for example, DBO in CB[7] has a fluorescence lifetime of 1 μs, which is the longest reported for an organic dye in a condensed phase of matter.

$$P = (n^2 - 1)/(n^2 - 2) \qquad (4.1)$$

Many different dyes have been encapsulated in CB[n], and a few notable examples are given below. A comprehensive list of dyes and their supramolecular complexes has been compiled by Nau.[254] Rhodamine 6G is a dye with extensive applications in life sciences, particularly in fluorescence microscopy. For example, the photophysical properties of CB[7]-encapsulated rhodamine 6G have been investigated.[59] Not only was the fluorescence lifetime extended but the brightness was also increased, and the rate of photobleaching was reduced. The improvement in properties allowed CB[7]-encapsulated rhodamine 6G to be used as an organic dye laser.[256] CB[7] has been used to encapsulate derivatives of cyanine dyes and thereby suppress their aggregation.[257] The extent of the aggregation could be controlled by addition of adamantylamine since it can displace the cyanine dyes. Fluorescent dyes conjugated to proteins are a valuable tool to visualise proteins of interest. However, poor photophysical properties can hamper such investigations. CB[7] has been used to improve the photophysical properties of a protein–dye complex.[258] For example, a triphenylmethane-based dye, Brilliant Green, has a very short lifetime (<1 ps) and a moderate affinity for bovine serum albumin. The addition of CB[7] increased the binding affinity by one order of magnitude, and the intensity of the fluorescence increased 300-fold, with an extended lifetime (1.5 ns). This cooperative binding system suggested that CB[7] encapsulation can improve the photophysical properties of dyes attached to proteins, which may have applications in bioimaging.

4.3. CB-Based Sensors

The highly specific host–guest chemistry of CB[n] can be used as the basis of sensing devices. For example, guest molecules can be sensed by the displacement of dye molecules; another strategy is to use fluorescent CB analogues that auto-report binding events. The applications of CB-functionalised electrodes to create highly specific and sensitive devices is also discussed.

4.3.1. *Fluorescent and other optical sensors*

Supramolecular displacement assays work by displacing an indicator from a macrocyclic host, thereby changing the fluorescence properties of the indicator. The groups of Anslyn[259] and Nau[25] have been key players in the development of such assays. They can either be a "switch-off" assay where the fluorescence decreases when the indicator is displaced by the analyte, such as the case of CB[7], where encapsulated dyes exhibit fluorescence enhancement. Alternatively, they can be "switch-on" assays where the fluorescence increases upon displacement. Nau, in particular, has pioneered the development of CB-based dye displacement assays.

CB[6]–dye complexes have been used as reporters in displacement assays to sense the binding of hydrocarbon gases in real time.[83] A key strength of this method is that the binding affinities of gas molecules can be measured in water since the CB[6]–dye complexes are soluble in pure water. Typically, salt is required to solubilise CB[6], but these salts outcompete weak binding molecules, such as hydrocarbons. Diaminohexane–dye conjugates form strong complexes with CB[6]; when gas molecules are bubbled into the solution containing the CB[6]–dye complex, the fluorescence continually dropped commensurate with the amount of gas in solution (Figure 4.8). The plateau region indicated the solubility limit of the hydrocarbon in water. By comparison of the plateaus of different hydrocarbons, the differences in binding affinity could be inferred. For example, isobutane showed three times higher affinity than the isomeric *n*-butane, presumably because it fills the cavity more effectively. Interestingly, the gas encapsulation is rapidly reversible, which is essential for gas-sensing applications. The system was restored by simply purging the solution with air.

A different dye displacement assay has been developed to identify quality (defined as high affinity and selectivity) binders for CB[7] and CB[8].[260] A dye (DMAT) that had a similar binding affinity for CB[7] and CB[8] was used (2.1×10^7 M^{-1} and 1.6×10^7 M^{-1}, respectively). DMAT is a water-soluble aryltropylium ion that absorbs UV light strongly due to intramolecular charge transfer. Interestingly, DMAT behaves differently in CB[7] and CB[8]: in CB[7] it undergoes a hyper-bathochromic shift, while in CB[8] it undergoes hypo-hypsochromic shift; this is manifested both spectroscopically and visually as a colour change in the solution. An equimolar mixture of DMAT, CB[7] and CB[8] resulted in an approximately equal amount of CB[7] and CB[8]–DMAT complexes. Next, a guest can be added, and its preference for CB[7] or CB[8] can be evaluated spectrophotometrically (and even by the naked eye) by virtue of the different solvatochromic behaviours. The method was validated with

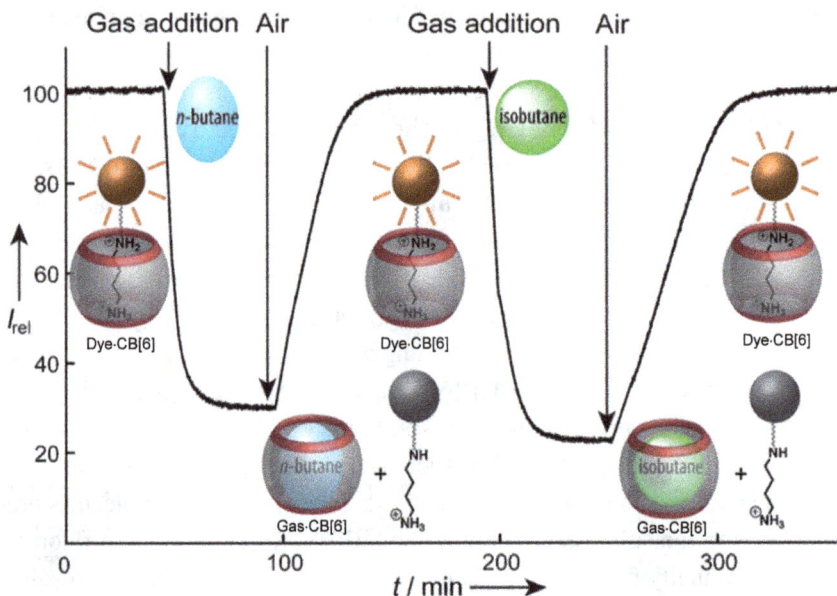

Figure 4.8. CB[6]–dye conjugates as a time-resolved hydrocarbon sensing system. (From Ref. 83)

known guests; furthermore, it was revealed that terpenes are a novel class of guest molecules for both CB[7] and CB[8]. A selection of CB[n] (and other molecular container) dye complexes have been used in colorimetric arrays to sense specific amines.[261] These arrays were able to discriminate between simple aliphatic and aromatic amines and give diagnostic patterns under both UV and visible light. Such arrays have also been used to detect illicit substances.[262]

CB analogues with inherent optical properties have been used as sensors for cancer-associated molecules, namely nitrosamines.[263] A fluorescent analogue of CB[6] acted as a selective sensor element (by virtue of its rigidity), and its sensing ability was augmented with a flexible, cross-reactive acyclic CB. When these two probes were applied in array format, selective identification of cancer-associated nitrosamines was achieved, even in the presence of interfering amines. A similar approach using only acyclic CBs was used recently to identify opiates and their metabolites.[264]

The dye displacement and florescence-sensing methods represent not only a time-resolved method of monitoring host–guest binding events, but also a high throughput alternative to more detailed affinity measurements, such as ITC and NMR.

4.3.2. *Supramolecular bioassays*

Nau *et al.* has pioneered the development of CB-based tandem enzyme assays for time-resolved measurement of enzyme activity (Figure 4.9). The main feature of tandem assays is that the product of the reaction (or substrate) and the dye molecules are competing for the host molecule, simultaneously. The major advantage of such assays is that they are "label-free", meaning that the enzyme activity is measured using real substrates and not fluorescently labelled analogues, which may have slightly different properties compared to those of the native substrates.

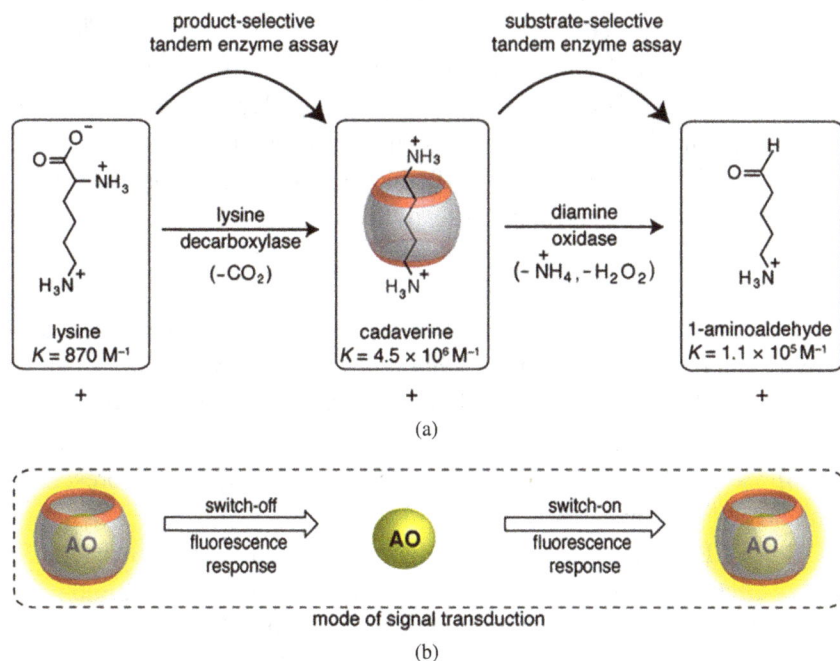

Figure 4.9. (a) A domino supramolecular tandem enzyme assay showing the product-selective assay and (b) the substrate-selective assay; the combination of these assays is termed a domino tandem assay. The chemo-sensing ensemble comprises the macrocyclic host CB[7] and the fluorescent dye, acridine orange. (From Ref. 25)

Decarboxylases are enzymes that remove carboxylates from substrates; one such enzyme is lysine decarboxylase which converts lysine to cadaverine; the latter can bind to CB[7] more favourably than lysine. The acridine orange-CB[7] complex (AO@CB[7]) was used to sense the product of this enzyme reaction; AO was used as a fluorescence indicator since AO exhibits enhanced fluorescence when encapsulated in CB[7].[265] When the enzymatic reaction occurs, the dye is displaced from CB[7] by the enzymatically produced cadaverine, thereby reducing the fluorescence output. The same principle has been applied to other classes of decarboxylases.[72,266]

The other mode of operation results in an enhancement of fluorescence output. When an oxidase reaction takes place with an ammonium substrate that is already bound to CB[7], the ammonium substrate is oxidised to an aldehyde. The affinity of the oxidised product to CB[7] is reduced enough so that it can be displaced by AO; once AO is encapsulated, the fluorescence increases in a time-resolved manner.[265] The two systems described above can be used in sequence to create a so-called "domino assay", where lysine is converted to cadaverine and then converted to aminoaldehyde; the fluorescence changes from on, to off, to on, respectively.[265] The sensing of specific biological molecules using a supramolecular assay is powerful because all that is required is a host and a dye molecule, whereas many biological assays require a different antibody or chemical probe for each substrate.

Using the same principle, but with a CB[6]-dye pair, acetylcholinase activity can be measured.[25] Acetylcholine had already been established as a guest for CB[6].[134,267,268] Acetylcholinase catalyses the hydrolysis of acetylcholine to choline; choline does not bind as strongly (100-fold less) to CB[6] and is displaced by DSMI, leading to an increase in fluorescence as the reaction progresses.

Since CB[7] has the ability to recognise *N*-terminal aromatic residues, it has been used to prepare an assay for the protease, thermolysin. Thermolysin cleaves at the *N*-terminal side of a hydrophobic residue (such as phenylalanine) in a peptide[269]; the AO@CB[7] complex was again employed as a reporter in this system. The *N*-terminal amino acid motif exposed by the protease reaction works as a guest to replace AO in CB[7]. As the enzyme reaction progresses, the more AO is displaced from CB[7] by the product of the enzyme reaction, the greater the reduction in fluorescence. Other classes of proteases such as exopeptidases, which cleave *N*-terminal residue of a peptide and trypsin can also be interrogated using this approach.[270]

The advantages of supramolecular tandem assays is that they can probe protease action on substrates that cannot be monitored using conventional methods, such as fluorogenic peptides. A more general strength of these assays is that they can be applied in a high throughput manner, such as microtiter plate so that multiple substrates can be analysed and compared in a single experiment to determine the enzyme's substrate specificity.[266]

A further application of these assays is the determination of enantiomeric excess (*ee*) of an amino acid. These assays take advantage of stereospecific enzymes that process only one enantiomer of the substrate and the product from the enzyme reaction displace the dye from the CB[7], thereby giving a measure of the amount of that particular enantiomer. For example, lysine decarboxylase only decarboxylates *l*-lysine to cadaverine, which can then displace a dye in CB[7].[266] Whereas, *d*-lysine is not processed. CB[8] complexes can also be used to sense chiral amino acids.[271] Some dye@CB[8] complexes have room for a second guest, such as aromatic chiral amino acids. Chiral amino acids themselves are not easily measured by circular dichroism (CD); however, when a supramolecular complex is formed including a chiral component, it becomes much more sensitive to CD due to increased size. The chiral ensembles could be detected in μM concentrations. Furthermore, the absorbance in the far-UV region (>310 nm) means that there is little interference from biological matrix effects. Such measurements can be coupled to an enzyme to monitor its activity.

In summary, CB[*n*] dye pairs represent a general system that can probe many different types of enzyme reaction, provided that the dye is a stronger binder than either the starting material or the product, but a weaker binder than the respective component (Table 4.1). Dye selection is, therefore, crucial for making dynamic and responsive systems. CBs are not the only supramolecular host to be used for this purpose, other supramolecular hosts have also been used for enzyme assays, details of which can be found in Nau's reviews.[25,254]

Table 4.1.   Dynamically responsive CB-based enzyme assays. The more strongly binding analyte that displaces the dye molecule is underlined. (Modified from Ref. 25)

Host–Dye	Enzyme	Substrate/Product	Ref.
CB[7]–DAP	Amino acid decarboxylase	amino acid/<u>biogenic amines</u>	72
CB[7]–AO	Lysine decarboxylase	lysine/<u>cadaverine</u>	265
	Diamine oxidase	<u>cadaverine</u>/aminoaldehyde	265
	Thermolysin	polypeptides/<u>dipeptides</u>	269
	Leucine aminopeptidase	<u>polypeptides</u>/amino acids	270
	Trypsin	polypeptides/<u>peptide fragments</u>	270
CB[6]–DSMI	Acetylcholinesterase	<u>acetylcholine</u>/choline	25
CB[8]–MDPP	Leucine aminopeptidase	polypeptides/<u>amino acids</u>	271

Notes: DAP = Dapoxyl; AO = Acridine Orange; DMSI = Hemicyanine; MDPP = Dimethylated peropyrenium.

4.3.3. *CB-functionalised electrodes*

Electrochemical sensors for chemicals and biochemicals show great promise for detection of analytes either in the field or at the point-of-care. Electrode-based assays possess many strengths including low cost, ease of fabrication and miniaturisation, and high sensitivity. Thus, it was envisioned that electrodes functionalised with CB[*n*] might serve as sensors for specific ions or molecules.

Since CB[6] has selectivity for specific cationic species, Kim *et al.* thought that CB[6] might be used as a component of an ion-selective electrode (ISE). The biggest problem they faced, to begin with, was the poor solubility of CB[6] in many common solvents, which prevented investigation of their host–guest chemistry in pure water. To address the poor solubility, they synthesised a new CB[6] derivative that was functionalised around the periphery. Specifically, star-CB[6] (CB*[6]) that has a cyclohexyl group fused to each glycoluril unit to make a functionalised "equator". CB*[6] is water soluble, which allowed the analysis of host–guest chemistry in pure water. Furthermore, CB*[6] was soluble in organic solvents as well, which facilitated the fabrication of membrane-based electrodes that could be used to sense for ions or cationic molecules.[134]

Acetylcholine (ACh⁺) is an important biomolecule to be assayed because of its role in the nervous system as a neurotransmitter. However, its lack of chromophore or electroactive group makes detection of ACh⁺ in the presence of other ions challenging. Conventional methods of assaying ACh⁺ rely on the immobilisation of acetylcholinesterase and detection of the enzymatic activity to ACh⁺. Therefore, a synthetic receptor that could detect acetylcholine directly may be very useful for diagnostics related to nervous system disorders. It was found that CB*[6] binds ACh⁺ strongly even in the presence of interfering ions.[134] Further improvements were made by using a different CB[6] analogue, perallyoxyCB[6], AOCB[6] that exhibited sensitivity for ACh⁺ at a concentration as low as 10^{-6} M^{-1}.[267] More importantly, the closely related molecule choline (Ch⁺) could only be detected at concentrations greater 10^{-4} M^{-1}. Such discrimination is significant since Ch⁺ is present at around 1000-fold higher concentration in physiological conditions. The improvements to the selectivity and detection of ACh⁺ were not due to any difference in the recognition properties, but instead a difference in solubility and membrane compatibility. AOCB[6] is soluble in alcoholic organic solvents such as methanol, which is compatible with the polymer membrane and helps impregnate the CB into the surface; however, it is essentially insoluble in water, which prevents it from washing away during the assay.

While functionalisation of the host molecule improved the sensor performance, it was still some way short of the best enzyme-based ACh⁺-based biosensor: Acetylcholinesterase immobilised on a carbon nanotube field effect transistor (FET) exhibits a limit of detection of 10^{-10} M^{-1}.[272] In an effort to improve the sensitivity of the CB-based system, the detection was carried out on an organic field-effect transistor (OFET, Figure 4.10).[273] OFETs are known to be very sensitive, easy to fabricate and work well in aqueous systems, but they do not have very much intrinsic selectivity. AOCB[6]-embedded OFETs combine the excellent selectivity of the host molecule with the high sensitivity of OFET devices to create a very sensitive and selective ACh⁺ sensor. Using this system, ACh⁺ could be detected at concentrations as low as 10^{-12} M^{-1}, six orders of magnitude lower than the previously reported ISE system; in addition, the OFET detected ACh⁺ with 100-fold better sensitivity than the best enzyme-based sensor does. This work demonstrated the great potential of CB[6]-based OFET devices as sensitive biosensors.

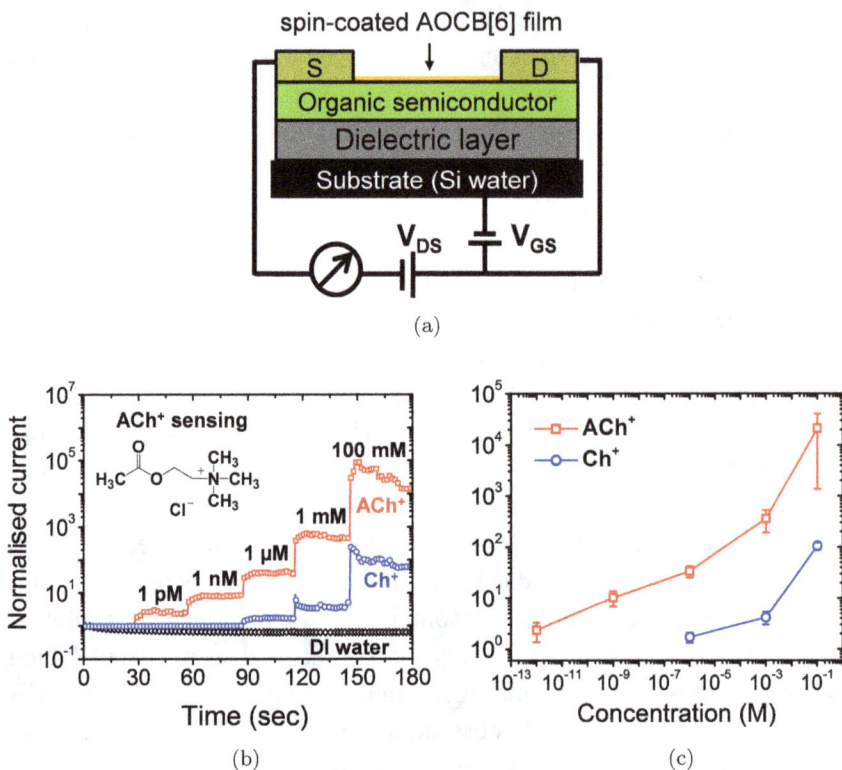

Figure 4.10. (a) Schematic representation of AOCB[6]-functionalised OFET-based sensors. (b) Real-time responses of AOCB[6]-functionalised OFET-based sensors towards varying concentrations of ACh$^+$, Ch$^+$ and DI water. (c) Statistical comparisons of sensing results. (From Ref. 273)

CB[7] has also been used in combination with OFET devices to make extremely sensitive sensors. Amphetamine-type stimulants (ATS) can be detected with picomolar sensitivity in biological fluids by preparing a OFET type sensor impregnated with allyloxy-functionalised CB[7].[274] The sensor could be prepared on a flexible substrate and connected to a smartphone application allowing for on-site use.[274–275]

Other important biomolecules can be sensed by use of CB[n]-functionalised electrodes. A CB[8]-functionalised glassy carbon electrode was used as a sensor that could determine tryptophan content in biological samples.[276] The CB[8] was introduced as a suspension with Naifon, which was added to improve the antifouling properties and passivate the surface.

Addition of tryptophan caused a peak shift in various electrochemical spectra. CB[8] is a good choice for this application, not only because of its recognition properties, but also its insolubility in water, which makes a stable surface. After showing that the tryptophan could be measured in the presence of interfering species at physiological concentrations, tryptophan was measured in human serum. CB[7] on glassy carbon electrodes has been used as part of glucose sensor.[277] Ferrocene was employed as a redox reporter, which was held to the surface using CB[7]. The ferrocene was used to sense the action of glucose oxidase in response to glucose.

4.4. Other Applications

4.4.1. *Waste water treatment*

The inclusion of guest molecules inside CB[n] in aqueous environments has led researchers to wonder if this property could be used for water remediation applications. Waste water from the textile industry often requires remediation to prevent dye molecules being introduced into the natural environment. The industry standard material for this purpose is activated charcoal; while charcoal is cheap and effective, it does not offer much selectivity. Behrend reported the ability of CB[6] to bind dye molecules such as Congo red and methylene blue in 1905. Buschmann, Schollmeyer *et al.*[278,279] demonstrated that CB[6] could bind other dyes, as well as other waste materials from the textile industry. Karcher *et al.*[45,280,281] have investigated CB[6] as a material for waste water remediation. They examined parameters such as pH and salt concentration on the dye sorption ability and concluded that although CB[6] is a potent sorbent for textile waste products; the dissolution of CB[6] in salt solutions and the aggregation of CB[6]-dye complexes were major drawbacks. To address these issues, solid-supported CB[6] may be required, but this may make the CB[6] materials too expensive to compete with activated charcoal. Nevertheless, a few advanced CB[6]-based applications have appeared for this purpose. Liu *et al.*[282] have used a CB[6]-based 2D-polymer film (the preparation of these films is discussed in Chapter 6) to adsorb methylene blue from aqueous solutions. While this polymer is a potent absorber of methylene blue, especially at the high pH where there are fewer cationic species, one of these issues is that the presence of salts inhibited the dye adsorption. In another example, CB[6]-impregnated into porous carbon materials could capture methylene blue or methyl orange.[283] The selectivity between these dyes can be tuned by adjusting the pH.

CB[6] may be useful for removal of hydrocarbons from aqueous solutions since simple hydrocarbons seem to prefer the CB cavity to the aqueous solution. For example, CB[6] can be impregnated into poly(urethane) sponges which can then be used for the removal of petroleum, diesel and soybean oil from both freshwater and seawater.[284] The material absorbed three times its weight in oil and could be reused by squeezing out the oil with mechanical pressure. The performance was slightly reduced in seawater because of the presence of salts. Acyclic CBs have also been shown to remove hydrocarbons from water.[107] The acyclic CB have sulphonates attached to them which makes them much more soluble than the cyclic CBs. The increase in solubility allows the acyclic material to outperform CB[8] in terms of hydrocarbon uptake. While these acyclic CBs may not be useful for bulk water remediation, they may have specialist uses such as separation or sequestration of specific hydrocarbons from a mixture.

4.4.2. *CB-anchored silica*

The anchoring of CB[*n*] to silica has been used to prepare materials for the chromatographic separations of small molecules. The first such material was disclosed in a patent by Kim *et al.*[285] Other studies have reported the grafting of perhydroxyCB[6] onto isothiocyanate-functionalised silica, which was then used as a hydrophilic interaction chromatography solid support.[286] This modified silica could separate molecules such as alkaloids, alkanes, aromatic hydrocarbons, alcohols, esters and ketones. The presence of salts in perhydroxyCB[6], an artefact of its preparation, interfered with the grafting of the CB to the silica. To address this, perallyloxyCB[6] was employed as a different reaction to immobilise the CB onto silica. Thiolated silica was irradiated with UV light in the presence of perallyoxyCB[6] to immobilise the CB through a thiol–ene reaction.[67] Although this method allowed a clean and successful immobilisation, not every CB[6] was accessible to small molecules.

Chapter 5

Supramolecular Systems Built with Cucurbiturils

5.1. Mechanically Interlocked Molecules

The mechanical bond is one of the signature motifs in supramolecular chemistry. Mechanically bonded molecules are not connected by bonds between the atoms, but instead interlocked in space.[4] Representative examples of such molecules include rotaxanes and catananes (Figure 5.1). Rotaxanes are dumbbell-shaped molecules that consist of a linear molecule (string) with a macrocycle threaded onto it (bead) with bulky stoppers at both end to stop the bead from unthreading. Pseudorotaxanes are similar structures, but they only consist of a thread and a bead; as a result, pseudorotaxanes can be disassembled. Catananes are chain-like structures that consist of interlocked molecules that cannot be separated without breaking a covalent bond. All of these structures can act as repeating units in polymeric structures. CB[n] are macrocycles that can act as beads in both rotaxane, pseudorotaxane and catanated structures.

5.1.1. *Simple CB-threaded rotaxanes*

The host–guest behaviour of CB[n] allows the threading of organic molecules containing specific motifs through the cavity. Simple alkyldiammoniums are an example of guests that can do this, which results in the

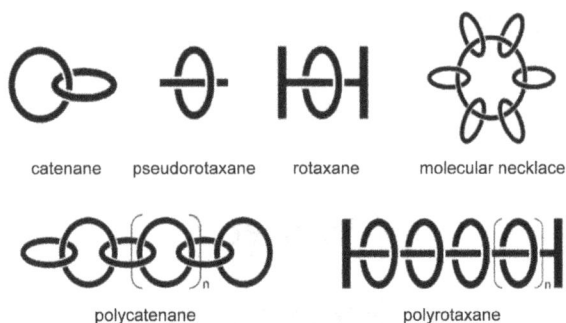

catenane   pseudorotaxane   rotaxane   molecular necklace

polycatenane   polyrotaxane

Figure 5.1.   Cartoon representations of various interlocked molecules.

formation of pseudorotaxane. Mock was the first to realise that CB[6] could be used in such a way (Figure 5.2(a)).[11] The portal can make ion–dipole interactions with ammoniums, and the alkyl chain threads through the hydrophobic cavity of the CB[6]. This molecule also exhibits pH-driven switching behaviour, which will be described later in this chapter. Rotaxanes can also be made from alkylammoniums, such as spermine.[287] When spermine and CB[6] are mixed in water, a pseudorotaxane is formed; the non-interacting amines at both termini can then react with electron-deficient aromatic-fluorides to add stoppers to complete the rotaxane. This is a simple, high-yielding route to simple CB[6] rotaxanes.[288] The first CB[7]-threaded rotaxanes and pseudorotaxanes were prepared using a similar strategy (Figure 5.2(b)).[288] Instead of using spermine as the axel, viologen was used. Depending on the functionality on the viologen, either an external complex or an inclusion complex was formed. Simple alkyl chains favoured the exclusion complex. However, if bulky stoppers were added, such as those used in the case of CB[6], then a pseudorotaxane was formed; CB[7] is big enough to "slip" over the stopper, but once threaded it is difficult to remove. A CB[7] pseudorotaxane can also be formed from a xylene flanked by pyridiniums groups, and these can be made into rotaxanes by adding an even bigger stopper. CB[8] rotaxanes have also been realised.[208] They were prepared using viologen to first thread CB[8] onto the axel, then using appended azide groups, a click reaction was performed to add bulky trityl-based stoppers. Despite being in a rotaxane, the threaded CB[8] retained a binding site for a second guest, such as the biologically important molecule, tryptamine.

CB[6]

(a)

CB[7]

(b)

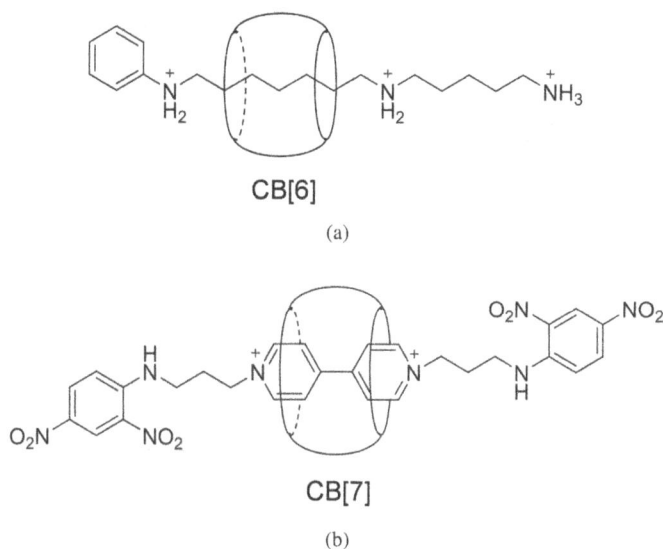

Figure 5.2. (a) A CB[6] pseudorotaxane and (b) a CB[7] rotaxane.

5.1.2. *Metal-directed self-assembly of CB-containing polyrotaxanes*

To translate CB[6] rotaxanes into polyrotaxanes, metals have been used as linkers that also dictate the geometry of the assembly. Metal cations make reasonably strong coordination complexes with appropriate organic ligands, and the metal centres often have preferred geometries, which allows rational design of the polyrotaxanes (Figure 5.3).

One-dimensional polyrotaxanes can be prepared from simple rotaxanes that have a metal-chelating pyridyl group as the stopper unit. Depending on the identity of the metal and the position of the *N* on the pyridyl ring different geometries of polyrotaxanes are generated. For example, when **PR44** is used and the metal is Cu^{2+}, a zig-zag 1D polyrotaxane is formed because of the square pyramidal geometry of the Cu^{2+}: and two adjacent sites on the Cu^{2+} are occupied by pyridines to give a *cis* geometry and the remaining sites are occupied by water molecules.[44] Different metals yield different geometries: Co^{2+} and Ni^{2+} with **PR44** give a square wave and a linear chain, respectively.[44] Silver ions have a high propensity to form two coordinate linear complexes, and this can be exploited to produce 1D polyrotaxanes. The pyridyl units coordinate with

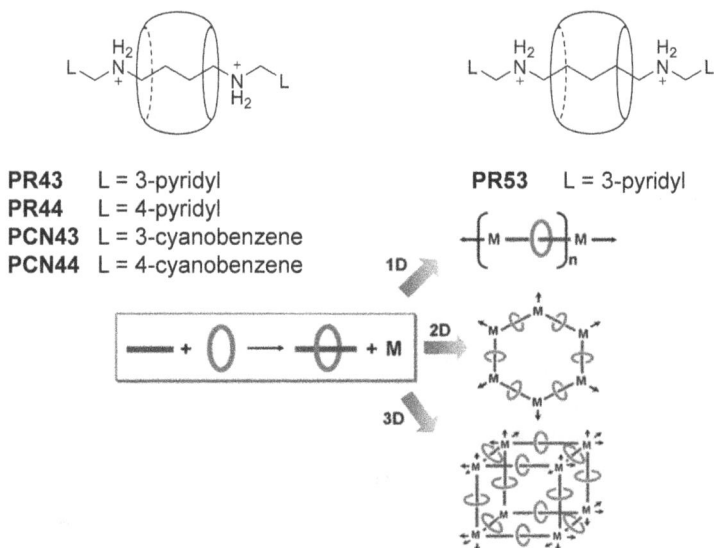

PR43 L = 3-pyridyl
PR44 L = 4-pyridyl
PCN43 L = 3-cyanobenzene
PCN44 L = 4-cyanobenzene

PR53 L = 3-pyridyl

Figure 5.3.   Structures of pseudorotaxanes and their use in the preparation of 1D, 2D and 3D structures.

the metal ion in a *trans-* geometry to give a straight-chain polymer. When the pyridyl N is at the 4-position treatment with Co^{2+} and Ni^{2+} yields zig-zag shaped polyrotaxanes, since they coordinate in a *cis* geometry.[17]

Helical 1D-polyrotaxanes can be generated from metal cations and **PR53**, which has a longer alkyl chain and than that of **PR44**.[289] Addition of the elongated dipyridyl to a solution of $AgNO_3$ yields a helical polyro-taxane. The helix, in which two rotaxanes and silver ions constitute a turn, extends along the *b*-axis of the crystal with a pitch of 17.9 Å. The coordination is in a linear arrangement. When Cd^{2+} is used as the linking metal, a different polyrotaxane is formed, and each turn of the helix is comprised of four rotaxanes and two different Cd^{2+} ions, leading to a pitch of 50.5 Å. One type of Cd^{2+} ions is coordinated with the pyridyl in a *trans*-fashion, and the other type of Cd^{2+} ions is *cis* coordinated.[17] Both these helices form a racemic mixture of left- and right-handed helices.

The identity of the metal counterion is also a key feature in determining the solid state structure. When **PR44** is formed from silver nitrate, a 2D polyrotaxane is formed.[290] The 2D network consists of large chair-shaped hexagons with a silver cation at each vertex, and each silver ion is complexed to three polyrotaxanes. In the solid state structure (Figure 5.4(a)), two independent 2D polyrotaxanes interpenetrate each other with full interlocking of the hexagons to form a polycatanated net. Changing the pyridyl position from 4 to 3 results in a different 2D poly-rotaxane.[17] The reaction of silver triflate with **PR43** yields a square meshed 2D-coordination polymer with CB[6] "beads" (Figure 5.4(b)). The triflate anions fill the interlayer space. In topological terms, these structures are (6,3) and (4,4) nets, respectively.[291]

Three-dimensional polyrotaxanes require slightly different design features. Larger lanthanide cations are used as the linking metals and aro-matic carboxylates are used as the ligands.[292] Cyano-stopped **PCN43** mixed with $Tb(NO_3)_2$ under hydrothermal conditions produces a 3D poly-rotaxane framework consisting of a binuclear Tb^{3+} and six pseudorotaxane units, each of which contains 3-phenylcarboxylate groups at the termini (Figure 5.4(c)). The resultant unit cell of the 3D-polyrotaxane has an inclined α-polonium topology, with the Tb^{3+} centres acting as six-connected nodes.[291] Like in the aforementioned examples, a simple change to the molecular structure affords totally different solid state struc-tures. Pseudorotaxanes that are terminated with a 4-cyanophenyl (**PCN44**) produces a 2D catenated structure; in this case, the Tb^{3+} behaves as a three-connected node.

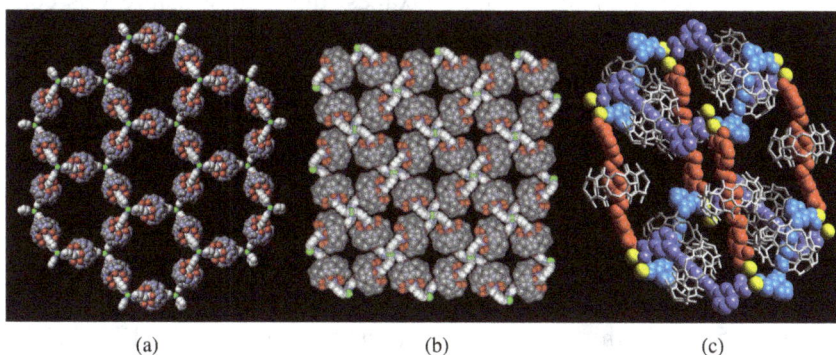

(a)   (b)   (c)

Figure 5.4. (a) Hexagonal 2D polyrotaxane net formed from **PR44** and Ag^+, (b) square grid 2D polyrotaxane net formed from **PR43** and Ag^+ (from Ref. 290), and (c) a 3D poly-rotaxane formed from Tb^{3+} and **PCN43**. (From Ref. 292)

The large open metal-polyrotaxanes described in this section may have applications as open framework materials.[293] Kim *et al.*[294] have investigated along this line by preparing a 2D-polyrotaxane network with large cavities and channels, thereby demonstrating a route to porous solids. The prepared material allows the size-selective exchange of the anions suggesting that molecular recognition applications may be possible. A further application of this chemistry is the incorporation of switchable rotaxanes into a framework structure as a route towards solid state supramolecular machines.[295]

5.1.3. *Molecular necklaces*

The above strategy can not only be used to construct linear polyrotaxanes but also cyclic polyrotaxanes that form discrete shapes, such as molecular necklaces. A molecular necklace is a cyclic, non-covalently linked molecule made from CB[6] pseudorotaxanes connected by ligand metal complexes. Pt complexes, which have square planar geometry, with *cis* vacant coordination sites can serve as angle connectors in the synthesis of squares and cages;[296] they can perform a similar role in the preparation of molecular necklaces.

Kim *et al.*[297] showed that when pseudorotaxane **PR44** was refluxed in the presence of Pt(ethylenediamine)(NO$_3$)$_2$, a molecular triangle was formed (Figure 5.5(a)). The Pt complexes are the vertices of the triangle, and are connected by the pseudorotaxanes; the final structure was confirmed by X-ray crystallography. Interestingly, each Pt cation is separated by 19.5 Å, which leaves very little space inside the interior of the triangle, suggesting that hydrophobic interactions between the exteriors of the CB[6] beads may play a role in stabilising these complexes.

Similar to the coordination polymers above, changing the position of the pyridyl nitrogen dictates a change in the supramolecular structure. Moving the pyridyl nitrogen from 4 to 3 (**PR44** to **PR43**) position results in the formation of a molecular square, under the same conditions (Figure 5.5(b)). In this case, the Pt cations are separated by 17.7 Å. The X-ray structure revealed a slightly distorted square (bond angle 77°, instead of 90°) reminiscent of a butterfly with a "wingspan" of 22 Å. An interesting temperature-dependent effect was observed in the formation of

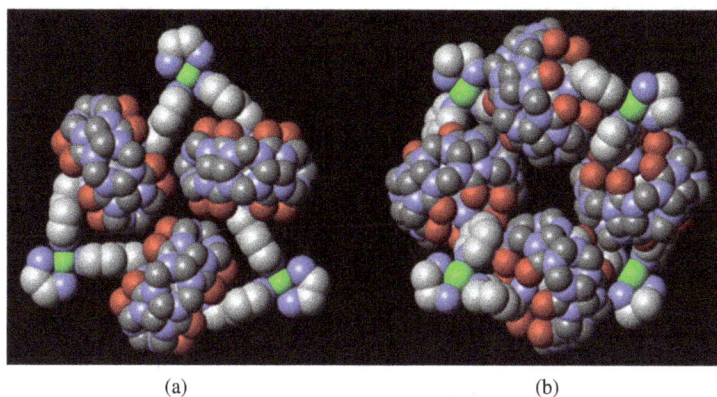

(a)   (b)

Figure 5.5.   (a) A triangular molecular necklace formed from **PR44** and Pt(En)(NO$_3$)$_2$ and (b) a square molecular necklace formed from **PR43** and Pt(En)(NO$_3$)$_2$. (From Ref. 18)

these two structures. At reflux, rotaxane **PR44** forms the triangle exclusively; however, at room temperature a 1:1 mixture of the triangle and square is observed. This suggests that there is competition between the entropically favoured triangle and the enthalpically favoured square at room temperature. At reflux, the strain (resulting from the tighter angles) is overcome and forces the formation of the entropically favoured product.[17]

Another approach to make square-shaped molecular necklaces is the so-called 2 + 2 approach,[298] where an L-shaped fragment is preformed and then dimerised by metal complexation to make the square. The L-shaped ligands had a phenyl or phenanthroline core with two diaminobutane arms, each of which was threaded with a CB[6], then stoppered with a pyridyl group. These components were then allowed to react with a metal such as Pt(en)(NO$_3$)$_2$ or Cu(NO$_3$)$_2$ to produce the molecular necklaces.

5.1.4. *CB-based organic polyrotaxanes*

In addition to metal coordinating polymers, organic-based polypseudorotaxanes have also been reported. An early example of such a material consisted of spermidine-CB[6]-based pseudorotaxanes linked by diamides.[299] However, the characterisation and application of this polymer was hampered by their poor solubility — they only seem to be soluble in sulfuric acid. In a different approach, a poly-spermidine polymer was connected through amide bonds, then the amide was reduced to the corresponding amine, and finally, the CB[6] beads were threaded onto the already formed polymer.[300] However, the threading was slow. Pseudo-polyrotaxanes containing CB[6] threaded onto a polyviologen polymer where the CB[6] beads occupy the alkyl chains in between the viologens have been described.[301] The ion–dipole interaction between the portal and the hydrophobic interactions between the cavity and the alkyl chain were responsible for the threading. In contrast to other pseudorotaxanes of this type, this polymer is water soluble.

A more fruitful way of producing polypseudorotaxanes is to thread them during the synthesis. An interesting example of organic polyrotaxanes has been reported by Steinke[302] that utilises a CB[6]-mediated rate acceleration of 1,3-dipolar azide–alkyne cycloaddition, as the linking chemistry. Such a system could be used to prepare "self-threading" polymers that were linked by CB[6]-occupied triazoles (Figure 5.6).[303] This method has proved to be a versatile approach to preparing many different polyrotaxanes, containing not only simple aliphatic or aromatic spacers but also groups with interesting properties such as porphyrins.[229] An important consideration is the length of the spacer unit, in order for efficient rate-accelerated elongation; CB[6] should have higher affinity for

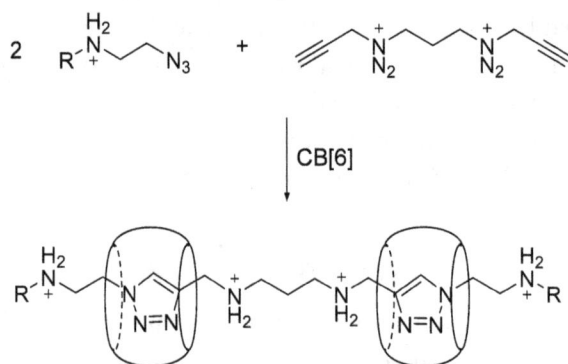

Figure 5.6.   The preparation of CB[6]-containing organic pseudorotaxane using CB[6] to facilitate the 1,3-dipolar cycloaddition.

the methylamino-amine and methylamino-azide than for the spacer unit. In structural terms, this means that the alkyl linker should be shorter than butyl, or longer than hexyl. Using this approach, polyrotaxanes of ~39000 molecular weight have been reported.[302]

More complex assemblies are possible that incorporate both CBs and cyclodextrins. Ternary complexes between CB[6], β-cyclodextrin and an alkyl amine can be formed.[304] These complexes exhibit positive cooperativity, β-cyclodextrin binds to alkyl chain 33-times stronger when CB[6] was present first. A quaternary complex can be formed with a three-station pseudorotaxane.[305] β or γ-cyclodextrin occupies the central station and two CB[6]s occupy the terminal stations. The CB[6] s can be removed by addition of a high-affinity guest, such as spermine.

Progress in the field of CB-based polyrotaxanes has been reinvigorated by the advent of so-called "cooperative capture synthesis".[306] In cooperative capture synthesis, other macrocyclic host molecules are used to template the CB-catalysed azide–alkyne cycloaddition reaction. The first examples of this rate acceleration used γ-cyclodextrin and CB[6], resulting in a rate acceleration with near-quantitative yield (Figure 5.7).[91] A limitation of this approach is the limited substrate specificity of γ-cyclodextrin. To address this limitation, Stoddart *et al.*[307] utilised pillar[5]arene[308] as the templating host. Pillar[5]arene moves more freely along a variety of molecular axels and can stabilise transition states that γ-cyclodextrin could not. These polyrotaxanes have been used as solid state fluorescent materials for supramolecular encryption.[309] The reader is directed to Stoddard's recent review for more details.[306]

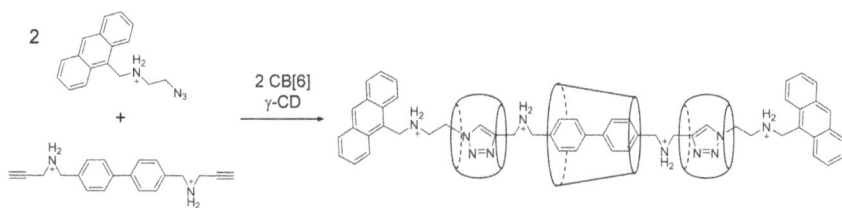

Figure 5.7.   Cooperative capture synthesis of hetero[4]rotaxanes using γ-CD and CB[6].

5.2. Supramolecular Assemblies Based on the Ternary Complexes of CB[8]

5.2.1. *Discrete complexes and assemblies built with CB[8]-stabilised charge-transfer complexes*

In Chapter 3, CB[8]'s ability to stabilise charge–transfer (CT) complexes was discussed; in particular, the complexes formed between methyl violo-gen derivatives and electron-rich naphthalene derivatives. Here, the dis-crete supramolecular assemblies that can be constructed using this chemistry will be described.

CB[8]-stabilised assemblies can be constructed using molecules that incorporate both a π-donor (D) and a π-acceptor (A) moiety separated by an appropriate linker are required (an A–D molecule). By varying the length and flexibility of the linker, various assemblies are possible (Figure 5.8). A long flexible linker favours a looped 1:1 host–guest complex, whereas a rigid and short linker, with dimethyldipyridyliumylethylene (DPE) as

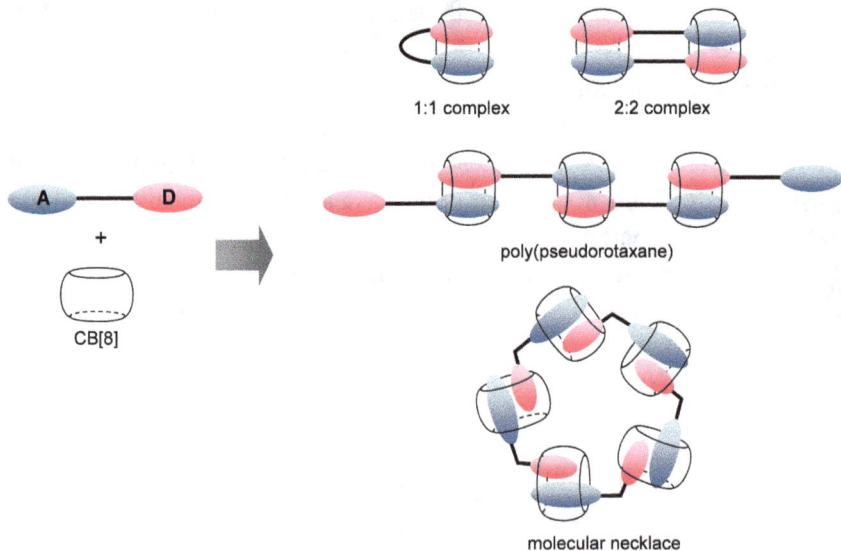

Figure 5.8. Possible supramolecular assemblies built with CB[8] and D–A molecules containing both electron-donor and electron-acceptor units connected by a suitable linker. (From Ref. 311)

the acceptor, favours a stacked 2:2 complex, or, on a gold surface, poly-(pseudo)rotaxanes can be formed by the layer-on-layer assembly.[310] A rigid linker with an appropriate angle between the D and A units can form a molecular necklace with CB[8] units as the beads.

An A–D (viologen–naphthalene, for example) molecule with a flexible linker can easily fold to bring the donor and acceptor molecules close together to form a CT-complex in the CB[8] cavity. Using a simple A–D molecule with a flexible four-atom link, a 1:1 intramolecular complex can be formed.[311] The formation of the complex can be observed by UV spectroscopy; the solution turns violet when the complex is formed. The structure can be further confirmed using NMR, specifically by the pulsed-field gradient (PFG) method. The complex thus formed was estimated to be 1.3 times larger than CB[8]. The complex can be switched from a binary to a ternary complex by addition of a higher affinity π-acceptor guest, such as DPE. The viologen is displaced, and a complex is formed between the naphthalene and the new guest.

Using an A–D molecule with a short, rigid linker such as a para *p*-xylene, and DPE as the acceptor leads to the formation of a 2:2 complex (or a (poly) pseudorotaxane).[310] The bulky linker prevents the intramolecular complex forming. The structure of this complex was elucidated using NMR; PFG–NMR revealed that the complex formed was 3.1 times bigger than CB[8].[311] By immobilising a DPE@CB[8] complex onto a gold surface a SAM could be formed onto which a polyrotaxane can be grown. This surface was treated with CB[8] and an A–D molecule. The growth of the polyrotaxane was monitored by FT–IR and SPR. Analysis of the gold surface after a day's polymerisation by AFM revealed 3.9 nm of growth. This polyrotaxane was "glued" together by CB[8]; this work, at the time, was a rare example of a non-covalent polymer grown on a solid surface.

Kim *et al.*[312] have used this supramolecular polymerisation method to construct cyclic oligomers, which resemble molecular necklaces. To achieve this, they designed an A–D that had a rigid but bent linker which not only prevents intramolecular complex formation but preorganises the monomer to form a pentagonal structure. Indeed, the molecular necklace structure is formed quantitatively. The structure was analysed by NMR, MS and X-ray crystallography. Using PFG–NMR, the size was

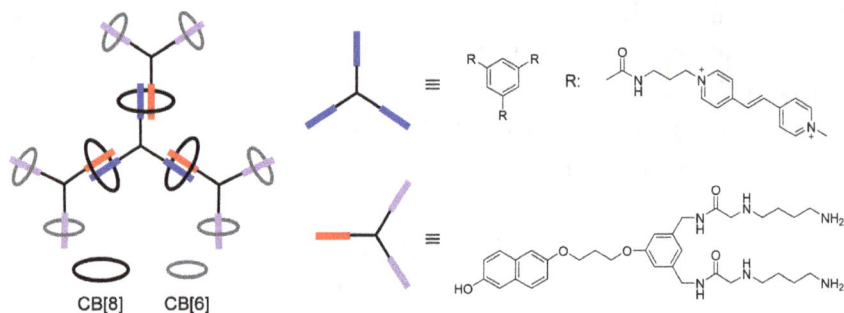

Figure 5.9.   A rotaxane dendrimer built with a CB[8]-stabilised CT interactions. (From Ref. 314)

estimated to be about ~9 times the size of CB[8]. The X-ray structure revealed that the necklace has a diameter of ~3.7 nm and a thickness of ~1.8 nm. The observed structure, despite some disorder, agreed well with the calculated structure. It may be possible to build molecular necklaces with different shapes and sizes by varying the angle and the length of the linker between the D–A. For example, a square CB[8] molecular necklace has been demonstrated recently, which uses a carbazole linker to make a ~90° connector.[313] In this case, the molecule is functionalised with methyl viologen on both sides, and assembly is induced by changing the pH.

The CB[8]-stabilised CT complexes can also be used to construct dendrimers (Figure 5.9).[314] The central molecule was shaped like a three-armed propeller terminated with dimethyldipyridyliumylethylene, this was connected to another propeller-shaped molecule in which, one arm had a hydroxynaphthalene and the other two arms consisted of diammoniumhexane which formed a complex with CB[6]. In total, a 13 component [10]pseudorotaxane was prepared.

5.2.2. *Vesicles built with supramolecular amphiphiles*

CB[8]-stabilised CT complexes can be used as building blocks for giant vesicles.[29] A 1:1:1 complex composed of dihydroxynaphthalene and methyl viologen with long alkyl chains inside CB[8] are supramolecular amphiphiles that spontaneously organise into giant vesicles. SEM, TEM

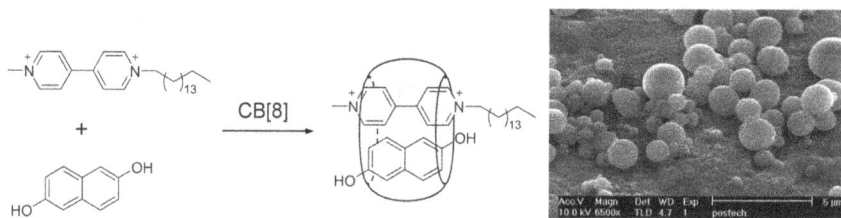

Figure 5.10. Formation of giant vesicles from a CB[8]-stabilised CT complex; an SEM image is shown on the right. (From Ref. 29)

and dynamic light scattering was used to analyse the formed vesicles (Figure 5.10). TEM revealed that the structures were hollow. Interestingly, these vesicles remain spherical and do not flatten when dried out on a surface, in the way that other synthetic vesicles do, as shown by SEM. Since the vesicles were constructed with a redox-responsive guest, the assemblies can be destroyed by treatment with oxidising or reducing agents. This work was the first example of a three component supramolecular amphiphile. Since then, other CB[8]-based vesicles and micelles have been reported, mostly in the context of drug delivery vehicles, which are discussed later (Chapter 7).

5.2.3. *Polymeric species built with CB[8]-stabilised CT complexes*

Since the initial demonstration of CB[8]-meditated polymerisation by the Kim group (described in the previous section) there has been much work on improving the degree of polymerisation. Generally speaking, there are two modes by which CB[8] can mediate the formation of polymeric materials: (i) small molecule-based polymerisation and (ii) block copolymerisation where homo- or heteropolymers with appended CB[8] binding groups are conjugated by CB[8].

Small molecule-based supramolecular polymers are constructed by the elongation of low-molecular weight monomers that are connected by CB[8]. There are several factors to consider when constructing a polymer in this manner. The first is the concentration and binding affinity of the components. For a high degree of polymerisation, both of these should be high.[315] The concentration is quite low due to the solubility of CB[8], but the high binding affinity compensates for that. The second consideration is how elongation can be promoted over the formation of macrocycles? In the previous section, it was shown that linker design is the key. Third is the binding mode; CB[8] is rare among supramolecular hosts in that they can incorporate two complimentary (or self-complementary) guests and this provides an opportunity to construct a variety of polymeric materials.

In the previous section, the formation of CB[8]-stabilised CT complexes with a covalently tethered A–D molecule was described. In the above examples, macrocycles (such as the molecular necklace) and simple linear polymers could be prepared. The main difference is in the linker design of the A–D molecule. Elongation of the polymer is favoured when the linker is rigid and linear so that it cannot fold back and form cyclic species. Another consideration is suppressing the formation of a 2:2 complex and promoting the polymer elongation. Using these concepts, Zhang *et al.*[27] prepared a water-soluble polymer. The monomer design incorporated four binding sites for CB[8], two acceptors and two donors (Figure 5.11). There are two viologens in the centre with terminal anthracene groups. The 2:2 complex is not formed because, by virtue of the short linker, the positively charged viologens would be too close to each other. A Job's plot analysis confirmed the 1:1 stoichiometry of the complex. A photoresponsive version of this polymer has been prepared by replacing the anthracene moieties with azobenzenes.[316] Upon irradiation with UV light, depolymerisation occurred. A self-sorted polymer that was controlled by both CB[7] and CB[8] has been reported.[317] The monomers consisted of a central dimethylamino-xyleneyl group with dimethylamino-naphthalene groups at both termini. CB[7] and CB[8] compete for all these sites. When only CB[7] or CB[8] is added, only simple pseudorotaxanes are formed. However, when one equivalent of both CB[7] and CB[8] are added a supramolecular polymer is formed, where CB[7] occupies the central xylene site, and CB[8] mediates the dimerisation between dimethylamino-naphthalenes, thereby elongating the polymer. Since the CB[8]-dinapthalene is not a strong complex ($K_a \sim 10^5$), the addition of extra CB[7] ($K_a \sim 10^6$ for the dimethylamino-naphthalene) can displace the CB[8] causing the polymer length to shorten.

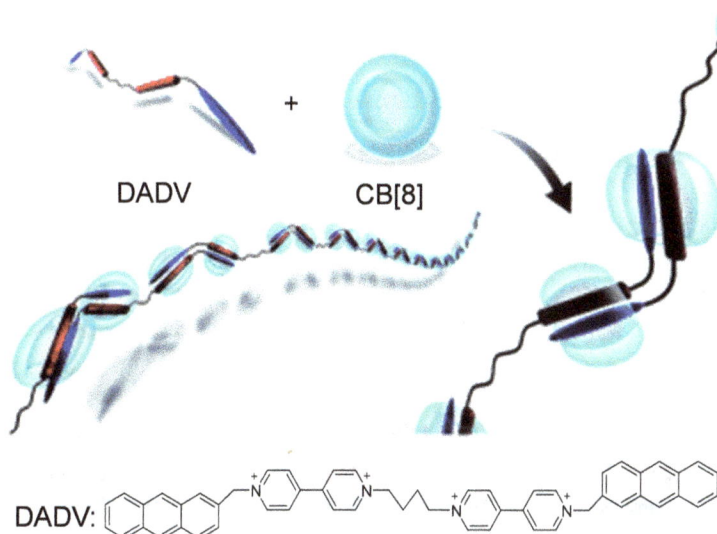

Figure 5.11.   Synthesis of supramolecular polymers using CB[8]. The monomer consists of two acceptor molecules (viologens) and two acceptor molecules (anthracenes) separated by a butyl chain. (From Ref. 27)

By the addition of more binding sites to the monomer, hyperbranched polymers, 2D or 3D networks can be prepared. By using monomers with three binding sites connected by a flexible linker, hyperbranched polymers can be realised.[318] Conversely, when the monomer molecule is rigid, the formation of a 2D network is observed.[319] These examples highlight the importance of the linker in the monomer design.

CB[8] can also be used to link together higher molecular weight species, such as conventional polymers, by complexing through their side chains or termini. Such a method allows the facile preparation of amphiphilic block copolymers and vesicles. CB[8]-mediated block copolymerisation was first demonstrated by Scherman *et al.*[28] (Figure 5.12). They demonstrated that naphthyl- and viologen-terminated polymers could be elongated into homopolymers or A–B diblock copolymers. Polymeric chain elongation mediated by CB[8] has been reported.[320] Low molecular weight polymers were synthesised by RAFT-polymerisation, and they possessed a naphthyl group at both termini.

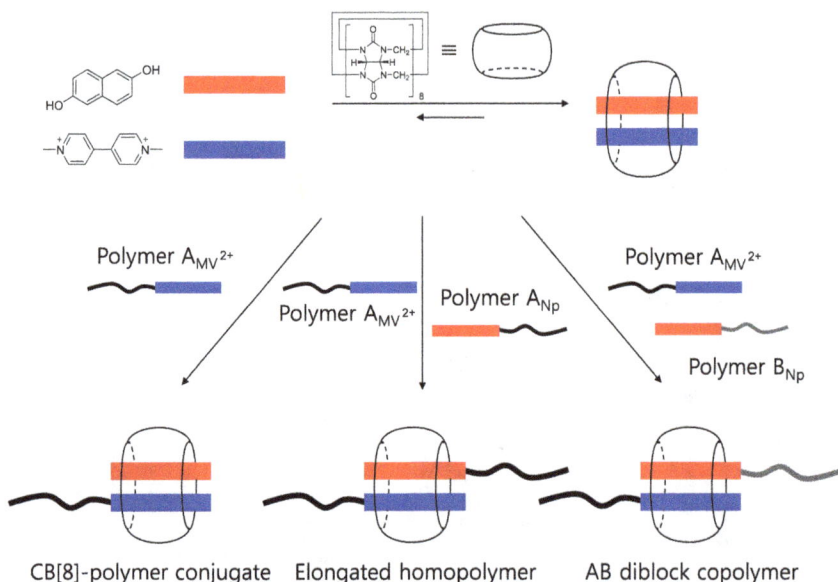

Figure 5.12.   CB[8]-mediated synthesis of polymers and diblock-copolymers.

Elongation was effected by the addition of viologen dimer and CB[8] in solution. The existence of the higher molecular weight copolymers was confirmed by viscometry techniques. Interestingly, these supramolecular polymers exhibited unusual lower critical solution temperature (LCST) behaviour in that it was strongly dependent on concentration, in contrast to covalent polymers where concentration has little effect on LCST. The polymeric species can self-assemble into vesicles and micelles, these are discussed later, in Chapter 7 in the context of drug delivery vehicles.

CB[8]-mediated modification of polymers can also occur on side chains.[321] This was achieved by using hydrophilic oligoethylene glycol polymer with naphthyl groups on the side chains. This side chain could then be modified with CB[8] and aliphatic methyl viologen derivatives. These polymers display the same redox responsiveness as other CB[8]-stabilised CT complexes, namely the side chain functionality can be disassembled and reassembled by redox chemistry. This approach has been used to prepare copolymers functionalised with biorecognition groups, such as lectins.[322]

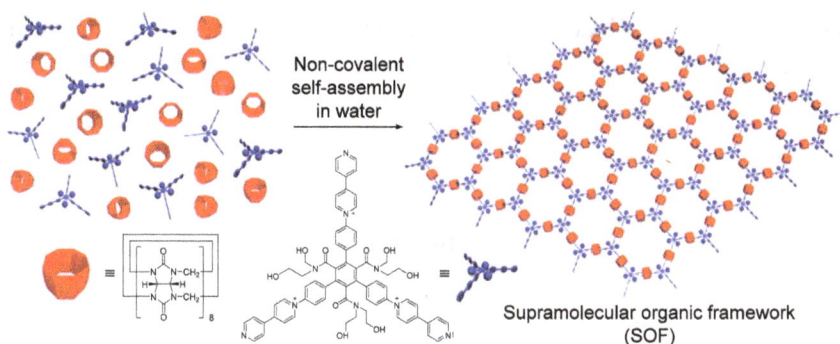

Figure 5.13. A single layer supramolecular organic framework (SOF) that is formed from dimerised bipyridinium derivatives in CB[8]. (From Ref. 96)

5.2.4. *Polymeric species built with CB[8]-stabilised homodimers*

Another aspect of CB[8] chemistry that can be exploited in polymerisation is the dimerisation of $MV^{+\bullet}$ inside CB[8]'s cavity. When MV^{2+} is reduced in the presence of CB[8] a 1:2 complex is formed exclusively.[245] This chemistry can be applied to the formation of polymeric CB[8]-based materials. Kaifer *et al.*[323] demonstrated that bipyridinium-functionalised dendrimers could be dimerised under redox control, thereby allowing redox switching of the molecular weight of the dendrimers. By preparing monomers that contain three bipyridinium, Li *et al.*[96] showed that a honeycomb-like structure could be obtained upon reduction of the bipyridinium and addition of CB[8] (Figure 5.13); a structure they termed a "supramolecular organic framework" (SOF). The same group extended SOFs into three dimensions; again the key was monomer design. The bipyridinium were organised around a tetrahedral trityl group, which after reduction in the presence of CB[8] formed a 3D-SOF.[102]

5.3. Molecular Machines and Switches

The high specificity binding exhibited by cucurbiturils can be modulated by several external stimuli, which has led to the development of stimuli-responsive machines and switches. In this section, the switchable behaviours of CB-complexes that are mediated by pH, light and redox chemistry are described. There are also many examples of CB-based switchable systems that are mediated by guest exchange. Such systems are described throughout this book, some of which are described in a recent account by Isaacs.[325]

5.3.1. *pH-controlled switches*

The strong preference that CB[*n*] has for positively charged amphiphiles can be used as the basis of a pH-responsive switch. Mock[11] reported that a CB[6] bead could migrate along a polyamine chain in response to pH change. At low pH, the CB[6] resides on a diprotonated diaminohexane (the optimal length for CB[6]), where the terminal amine is an aniline. When the pH is increased, the aniline becomes deprotonated first, and this causes the CB[6] to move down the chain to a protonated site. Kim *et al.*[326] used this principle to prepare a switch with a fluorescent output (Figure 5.14). In this case, the aniline portion was part of a dye molecule. When the CB[6] was adjacent to the dye, then the fluorescence is switched on; when the pH is increased, and the CB[6] moves down the chain, the fluorescence switches off. A ternary complex consisting of methylated β-cyclodextrin, CB[6] and alkylammonium chain also exhibits

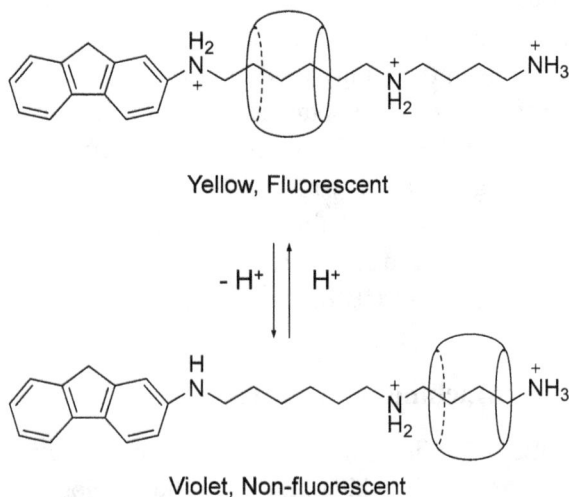

Yellow, Fluorescent

$- H^+$ ⇅ H^+

Violet, Non-fluorescent

Figure 5.14.   A pH-based molecular switch that controls the motion of CB[6]. The CB[6] motion is accompanied by quenching of the fluorescence since both are controlled by pH.

pH-switchable behaviour.[327] The two macrocycles prefer to interact with each other at high pH, and at low pH they are separate. pH-switchable CB[7]-based pseudorotaxanes has also been described.[328] A pseudorotaxane was prepared which had a central viologen flanked by alkylcarboxylic acids. When the pH is acidic, the CB[7] sits on the alkyl chain between the carboxylic acid and the viologen. When the pH is increased, the CB[7] moves to occupy the viologen station, because of the electrostatic repulsion between the CB[7] portal and the carboxylate. A dual stimuli CB[6] switch has also been prepared.[329] The movement away from the first station required only pH, but the reverse reaction required reversal of the pH change and an increase in temperature.

A CB[6]-based, pH-responsive nanovalve has been prepared by the groups of Zink and Stoddart.[74] They prepared a CB[6]-based rotaxane on the surface of mesoporous silica inside which molecules could be stored. The rotaxane serves as a "valve" between the mesoporous silica and the solution. When the pH was low, the CB[6] capped the mesoporous silica, thereby trapping the molecules inside. However, when the pH was increased, the CB[6] was removed, and the molecules inside were released (Figure 5.15). Such systems may be useful for drug delivery; however, for effective release of the molecules the pH had to be around 10, which is not biocompatible. The same researchers later addressed this limitation by preparing a similar system that released its cargo under mildly acidic conditions.[330]

They achieved this by using the same principle as Mock and Kim had previously demonstrated: by taking advantage of the pK_a differences

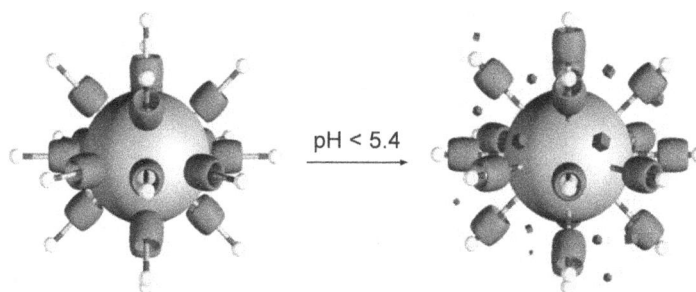

Figure 5.15.   pH-controlled nanovalves activated by low pH. (From Ref. 330)

between aliphatic and aromatic amines and CB[6]'s preference for being in-between two protonated amines six atoms apart, the CB[6] could be made to migrate along the chain by lowering the pH.

5.3.2. Light-controlled switches

A CB[7]-based photoswitchable host–guest system has been reported.[331] A cinnamide derivative can be switched from *cis* to *trans* by UV light (300 nm). The *trans*-isomer can form a complex with CB[7] ($K_a \sim 10^4$), whereas the *cis* complex does not; the process can be reversed by irradiating at 254 nm. This chemistry can be used to make a light-gated CB[7] nanovalve.[332] Similar to the CB[6] nanovalves described above, the CB[7] rotaxane is prepared on the surface of mesoporous silica. The CB[7] acts as a valve, and after irradiation with UV light, the cinnamide derivative switches its conformation and causes the CB[7] to dissociate, thereby opening the valve.

The well-documented photo-isomerisation exhibited by azobenzene can be used to prepare CB[8]-based light-responsive systems. The azobenzene *trans*-isomer can be switched to the *cis*-isomer by UV light, and the reverse reaction is possible either thermally or exposure to visible light. Scherman *et al.*[333] demonstrated that *trans*-azobenzene could form a ternary complex with MV^{2+}@CB[8]; *cis*-azobenzene does not form a complex because it is too bulky. Upon irradiation with UV light, the encapsulated azobenzene isomerises to the *cis*-isomer which is then forced out of the complex causing it to dissociate. The methyl viologen also provides a means to control complexation by redox chemistry (see below). The orthogonal switching ability of this complex allowed the preparation of rewritable memory device.[333] The azobenzenes can also be used to prepare a light-driven axle with CB[7] as the wheel.[334] The pseudorotaxane axel consisted of a central viologen unit with an azobenzene on both sides. In the ground state, each *trans*-azobenzene is encapsulated by a CB[7]. Upon irradiation to the *cis*-isomer, the CB[7] moves to central viologen station; however, since there is only one viologen, the other CB[7] is displaced into solution (Figure 5.16). This system allows control of both the molecular motion and the stoichiometry of the pseudorotaxane by light.

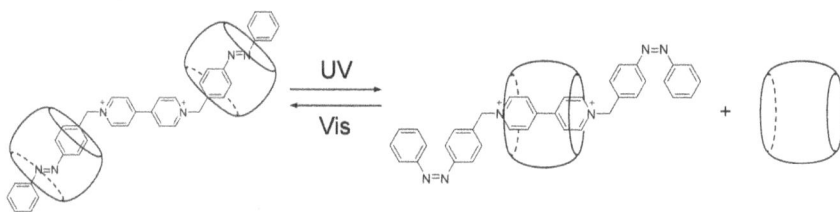

Figure 5.16. A light-activated molecular switch that releases CB[7]. (From Ref. 334)

5.3.3. Redox-controlled switches

There are a variety of redox-responsive CB[n] guests that give rise to CB[n] complexes that change in response to redox chemistry. Metallocenes, such as ferrocene, are tight binding CB[7] guests with high affinity and high specificity because of the size complementarity between the cavity and their hydrophobic core. The binding affinity can be increased by addition of cationic moieties to the hydrophobic core. Ferrocene (Fc) can undergo a single electron oxidation to become fer-rocenium, which has a positively charged core and has a reduced binding affinity for CB[7]. Kaifer et al.[335] have sought to reverse the complex formation of CB[7]-Fc by electrochemical oxidation of the Fc in the complex. When combined with deprotonation of the cationic groups, the stability of the complex can be reduced substantially ($\Delta\Delta G° = 6.5$ kcal mol^{-1}).[336] An electrochemically switchable pseudorotaxane based on CB[7] and Fc chemistry has been prepared.[65] The "axel" consisted of a p-xylene terminated with a trimethylamine ferrocene on both sides. In the ferrocene state, the CB[7] sits on the Fc; upon oxidation to ferroce-nium, the CB[7] migrates to the xylene and sits in-between the two amines. The process is reversible upon application of a reduction poten-tial. The reversible oxidation of the CB[7]–Fc has been used as a way to modulate the binding of a bulk material[95]; namely to detach supramo-lecular Velcro (see Chapter 8.5).

Another class of CB[n] guests that are redox-active are the viologens, such as methyl viologen (MV^{2+}). Viologens undergo two successive one-electron reductions form MV^{2+} to MV$^{+\bullet}$ to MV^0. As was shown in the previous chapter, CB[8] promotes the dimerisation of MV$^{+\bullet}$ in the CB[8]

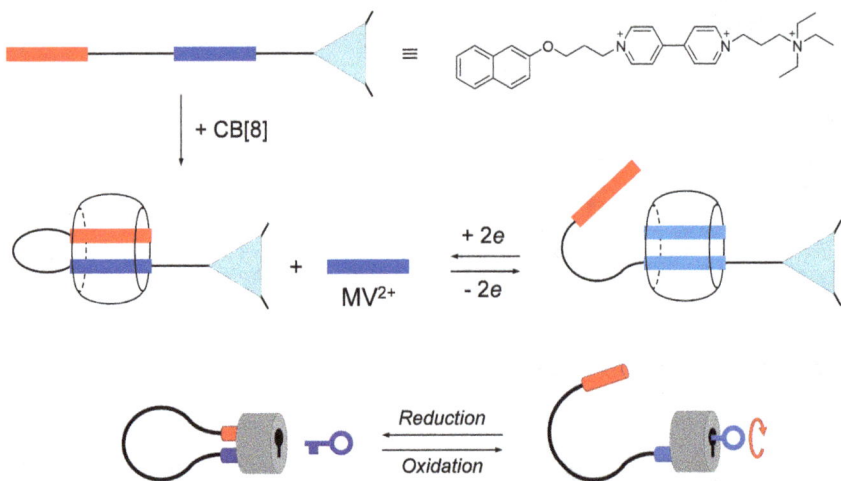

Figure 5.17.   A redox-driven molecular loop-lock with a redox-activated molecular key. (From Ref. 56)

cavity.[245,337] By exploiting this chemistry, CB[8] complexes wherein viologens act as a π-acceptor, can be switched from heterodimers to homodimer (or *vice versa*) by redox stimuli.

When the MV^{2+} and hydroxynaphthalene (HN) are tethered to each other, a redox-responsive molecular machine can be created: a molecular loop-lock (Figure 5.17).[56] The lock consisted of MV^{2+}, a short linker, HN and a bulky stopper attached to the MV^{2+}. The key was free MV^{2+}. However, it cannot be used until both itself and the complex are electrochemically reduced. The loop can be locked again by reoxidising with the introduction of oxygen. Such a system may be thought of as a "safeguarded lock" because it requires both a "key" and an electrochemical reduction to activate it. A similar system where the key is built in is also possible.[311] In this case, the chain consists of two viologen (V^{2+}) units and one HN unit. When CB[8] is added, the non-terminal V^{2+} and HN form an intramolecular CT complex. After both V^{2+} units are reduced to $V^{+\bullet}$, they form a dimer inside CB[8] and the HN moiety is displaced. This work was extended to make a three-way molecular switch based on CB[8], TTF and MV^{2+}.[247]

5.4. Self-Sorting Systems

A host–guest system is one where a host molecule forms a specific complex with a guest molecule, under a given set of conditions. A self-sorting system adds a level of complexity, where multiple, well-defined, host–guest complexes are present in the same system, under the same conditions; each host–guest complex forms with high fidelity with no cross reactivity. The CB[n] family possess several features which make them ideal candidates to prepare self-sorting systems. The high affinity and high selectivity of the host–guest pairs allows the formation of specific host–guest complexes in the presence of others (so-called social self-sorting), and the slow dissociation constants enables study of these systems by NMR. Indeed, self-sorted systems can be identified by the NMR spectrum. The NMR signature of a self-sorted system is that the spectrum of the whole system is equal to the superimposition of the spectra of the individual complexes.

The first indication that CBs may engage in self-sorted behaviour was discovered by Isaacs[123] during their investigations into acyclic CB[n] type molecular clips that consist of a glycoluril dimer flanked by an aromatic wall on each side. These molecular clips individually undergo homodimerisation in $CDCl_3$, mediated by H-bonding and π–π interactions from the aromatic walls. This work was extended to an aqueous solution where a 12-component system was prepared.[54] The system included the aforementioned molecular clips, CB[6], CB[8], β-cyclodextrin, an ionophore and a metal complex, and a guest molecule for each host. Despite the high potential for crossover complexes, the ensemble formed only a single set of complexes. These results were significant for two reasons; firstly because it demonstrated that social self-sorting could occur using synthetic receptors in aqueous systems. Previously it was thought that such self-sorting would be difficult in water because imprecise ionic and hydrophobic interactions dominate in water, whereas H-bonds are much more specific and prevalent in organic solvents. Secondly, it suggested that the specificity and affinity of these host–guest pairs were higher than previously thought.

The high selectivity that underlies the self-sorting behaviour can be attributed to the constrictive binding of CB[n] host–guest binding. The

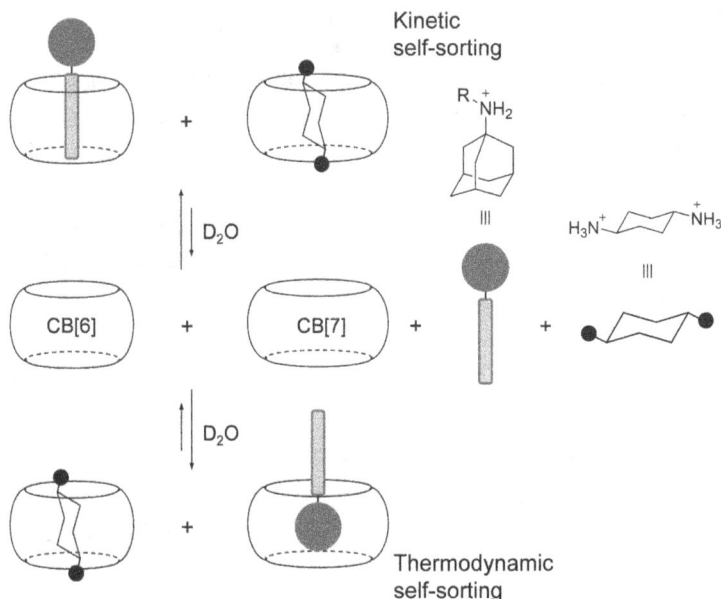

Figure 5.18. Self-sorting of the kinetic and thermodynamic complexes. Initially, the kinetic complexes form, but over a long period of time the guests switch to give the thermodynamic product. (From Ref. 66)

curvature of the glycoluril units means that the portal is narrower than the cavity. Therefore, some guests can form thermodynamically very stable complexes, but the rate of association is very slow; the converse of this is that rate of dissociation is also very slow leading to very tight binding reminiscent of biological systems. An example of such a system consisted of two hosts (CB[6] and CB[7]) and selected guests that could bind to both them (adamantaneammonium with an alkyl chain, and 1,4-diaminocyclohexane) (Figure 5.18).[66] The cyclohexane molecule can form inclusion complexes with both CBs, although the binding association kinetics are much slower with CB[6] than CB[7]. Similarly, the alkyl chain of the adamantane molecule can form an inclusion complex with CB[6], and the bulky adamantane portion of the molecule forms an inclusion complex with CB[7], but at a slower rate. When a solution of CB[6] and CB[7] is mixed with both these guests, they form a kinetically self-sorted mixture: the diaminocyclohexane@CB[7] and the alkyl

ammonium@CB[6]. However, over time (56 days) the thermodynamically favourable complexes emerge; the Ad@CB[7] and the cyclohex-
ane@CB[6]. The kinetic self-sorting system can be compromised by
elongating or shortening the alkyl binding epitope.

Short dipeptides have been shown to self-sort with CB[6] and CB[7].
Lys–Tyr prefers to bind to CB[6], whereas Tyr–Lys prefers to bind to
CB[7]; these complexes socially self-sort in aqueous solution.[338] The origin of the selectivity is that CB[7] has a strong preference for *N*-terminal
aromatic amino acid residues, whereas CB[6] binds to the side chain of
the lysine.

The kinetic and thermodynamic preferences of CB-based self-sorting
systems is also manifested in pseudorotaxanes; such processes are often
termed integrative self-sorting. For example, a three-station pseudorotaxane
consisting of tetra-ammonium alkyl thread that can hold three CBs exhibits
self-sorting behaviour (Figure 5.19).[79] When a mixture containing equal
amounts of CB[7] and CB[6] was added to a solution of the alkyl ammonium

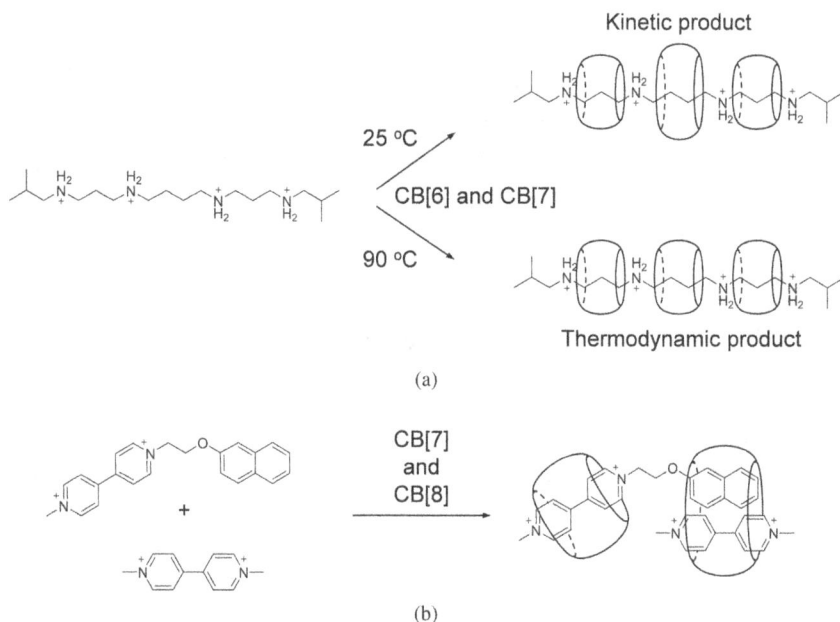

Figure 5.19.   (a) Self-sorted four component system and (b) self-sorted pseudorotaxanes.

at room temperature, the kinetic product, CB[6]–CB[7]–CB[6] is formed exclusively, since CB[7] has a lower barrier for association compared to that of CB[6]. However, at 90°C the thermodynamic product, CB[6]–CB[6]–CB[6], is formed because of the stronger binding affinity (Figure 5.19(a)). Schalley *et al.*[87] have extended the idea of integrative self-sorting to more complex pseudorotaxanes. Using an A–D type molecule, methyl viologen, CB[7] and CB[8] they demonstrated social self-sorting. Both CB[7] and CB[8] prefer to bind to the A–D molecule, but when both CBs are added, then a single complex is formed with high fidelity (Figure 5.19(b)). They next sought to demonstrate self-sorting using two axles (D–A–D and A–D–A). After the addition of three equivalents of CB[8], self-sorting leads to the only two isomeric complexes formed. These principles have been extended into longer chains that could be controlled by external stimuli such as the stoichiometry of CB[*n*], the presence of extra guest molecules and redox chemistry.[100] This six-station linear axel can either accept or expel CBs, with resulting changes to its geometry, in response to stimuli. The complexity of this system was reminiscent of a signalling cascade from a biological system. While the self-sorting behaviour demonstrated was high fidelity, it was also very sensitive to stoichiometric imbalances, which is a departure from how natural systems behave; nature is typically much more tolerant of stoichiometry. With regard to designing abiotic signalling cascades, such sensitivity could be considered either as a positive or a negative depending on the application of the system. Another area where this system differs from nature is that each state is the thermodynamic minimum, whereas nature frequently operates far from equilibrium.

5.5. CB Metal Complexes and Coordination Polymers

The CB family can interact with cationic metal ions to form exclusion complexes. The basis of these complexes is the highly polarised ureidyl portal which can make ion–dipole interactions easily. The carbonyls are close together, and so collectively they make reasonably strong interactions with metal cations to make complexes and coordination polymers. Some examples are presented here, but the reader is directed to a review by Tao[339] for a more complete look at the field.

CB[6] has a portal radius of 3.3–3.4 Å, so only Cs^+ is a large enough cation to fit over the entire portal. As such, Cs^+ ions form a bowl with

CB[6], where the Cs$^+$ interacts with four of the six carbonyls. However, when a coordinating organic guest, tetrahydrofuran (THF), is included inside the CB[6], all six carbonyls interact with the Cs$^+$; the oxygen from the THF seems to interact with Cs$^+$, which draws the metal ion closer to the portal.[183] A Na$^+$-lidded CB[6] capsule was demonstrated by Kim *et al.*[16]; again the formation of the metal CB complex was promoted by an encapsulated THF molecule in the cavity. This complex formation was reversible by addition of TFA, which displaced the guest.[47] CB[6] can form complexes with various heavy metals with varying structures. Fedin *et al.*[340] have investigated and categorised these into three groups. The first are 1:1 complexes where CB[6] binds one metal cation. In this case, the CB[6] acts as a bidentate ligand with several Ln^{3+}. The second are sandwich-type complexes, consisting of three CB[6]s bridged by two Sm^{3+} ions. The third type are 2:2 complexes where the CB[6]s are bridged by two Ln^{3+} cations; for example, Ce3+ forms complexes with CB[6] in this manner.[340]

CB[n] can also form coordination polymers with alkali metals. An early example reported that two Rb$^+$ ions form a bridge between CB[6] units, and this structure extends to form a 1D polymer.[341] Moreover, crystal packs together in a honeycomb arrangement to yield an open-channel structure (Figure 5.20). Each column is offset by one half of the repeating

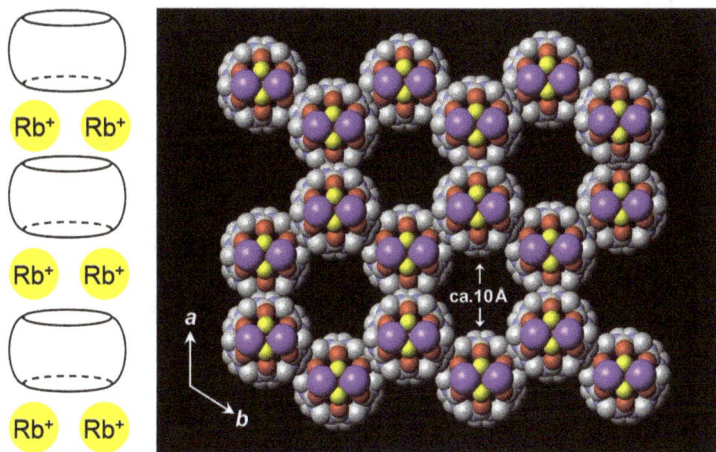

Figure 5.20. Rb-bridged CB[6] columns stacked together into a 2D-honeycomb. (Modified from Ref. 341)

unit which results in the formation of channels in the structure. CB[5] has a rich metal coordination chemistry due to its high proximal concentration of carbonyls and can form various coordination polymers. For example, CB[5] can form various coordination polymers with Sm^{3+} in the presence of different metals, counter ions and additives.[342]

The materials described here are generally not stable enough to form porous materials. However, related networked materials have taken advantage of CB[6]'s affinity for Cs^+ to make a molecular sieve for radioactive Cs^+ ions.[343] In the next chapter, more robust porous CB based materials will be discussed.

Chapter 6

Application of Cucurbiturils in Materials Science

6.1. Porous Materials

In Chapter 3, the ability of some CB[n] to capture gas molecules and specific cations was described. In this section, the translation of CB[6] into a solid state porous material endowed with host–guest chemistry of CB[6] is described.

Crystalline porous materials such as zeolites, metal organic frameworks (MOFs), covalent organic frameworks (COFs), and organic porous solids have great potential as gas sorption materials. The key advantages of porous material are that they have large surface areas and chemical functionalisation can tune their sorption preferences. However, current materials are sensitive to water or have low thermal stability, which limits their utility. CB[n], on the other hand, have exceptional thermal stability and are made from very cheap organic starting materials. As such, CB[n] represents a promising organic porous material for gas sorption.

DecamethylCB[5] was the first CB compound to be utilised for gas sorption. Solid, activated decamethylCB[5] could absorb CO_2 (26 mL g^{-1}) and N_2O (40 mL g^{-1}) efficiently.[51] Since CB[6] has exceptional thermal and chemical stability, Kim *et al.* exploited it as a solid state gas sorption material for acetylene[75] and CO_2.[344] Typically, CB[6] synthesised by the original Berhend method produces a crystalline polymorph that loses its crystallinity upon exposure to air. However, CB[6] recrystallised from HCl is an exceptionally stable material with permanent porosity. Even after complete removal of water and HCl in the lattice by heating under reduced pressure, it maintains its framework and is exceptionally thermally and chemically stable.[75] The X-ray crystal structure of porous CB[6] revealed a honeycomb-like structure with 1D channels along the *c*-axis, generated by a hexagonal arrangement of CB[6] molecules (Figures 6.1(a)–6.1(d)). Porous CB[6] has an acetylene storage density of 0.087 g cm^{-3} (under standard conditions) which is equivalent to the density of acetylene at 8.3 MPa at room temperature, and almost 42 times greater than the compression limit for the safe storage of acetylene (0.2 MPa). These values, at the time, were the highest reported for an organic porous material. Porous CB[6] also has a remarkable capacity to bind CO_2 (45 g cm^{-3} at 298 K) and shows higher selectivity for CO_2 over CO than other organic materials or most MOFs; the reason for this is the very high enthalpy of CO_2 adsorption (7.89 kcal mol^{-1}). Single-crystal X-ray analysis gave some insight into the origin of the high enthalpy of adsorption. It was found that CB[6] has three CO_2 adsorption sites; CO_2 binds not only to the cavity but also to the portals through hydrogen bonding and ion–quadrupole interactions (Figures 6.1(e)–6.1(h)). Such materials may be useful for the capture of CO_2 from flute gases in various industries. Later, other crystalline polymorphs of porous CB[6], one of which had superior CO_2 uptake capacity, were prepared.[345]

Porous CB[6] is also an effective conductor of protons in a highly anisotropic manner. The conductivity behaviour can be modulated by the nature and amount of acid molecules present in the channels. Single-crystal measurements revealed that proton conduction occurred along the direction of the channels. At the time, this material exhibited the highest anisotropic proton conduction of known conducting materials; the conductivity value (2.4 × 10^{-2} S cm^{-1} along the *c*-axis) is just one order of magnitude lower than well-established proton conductors such as Nafion.

Porous CB[6] can also conduct Li$^+$ ions.[346] The conductivity was measured to be ~10^{-4} S cm^{-1} along the *c*-axis, and this material exhibited good performance even after several high temperature cycles. This is a rare example of a porous organic solid that can conduct Li$^+$ ions.

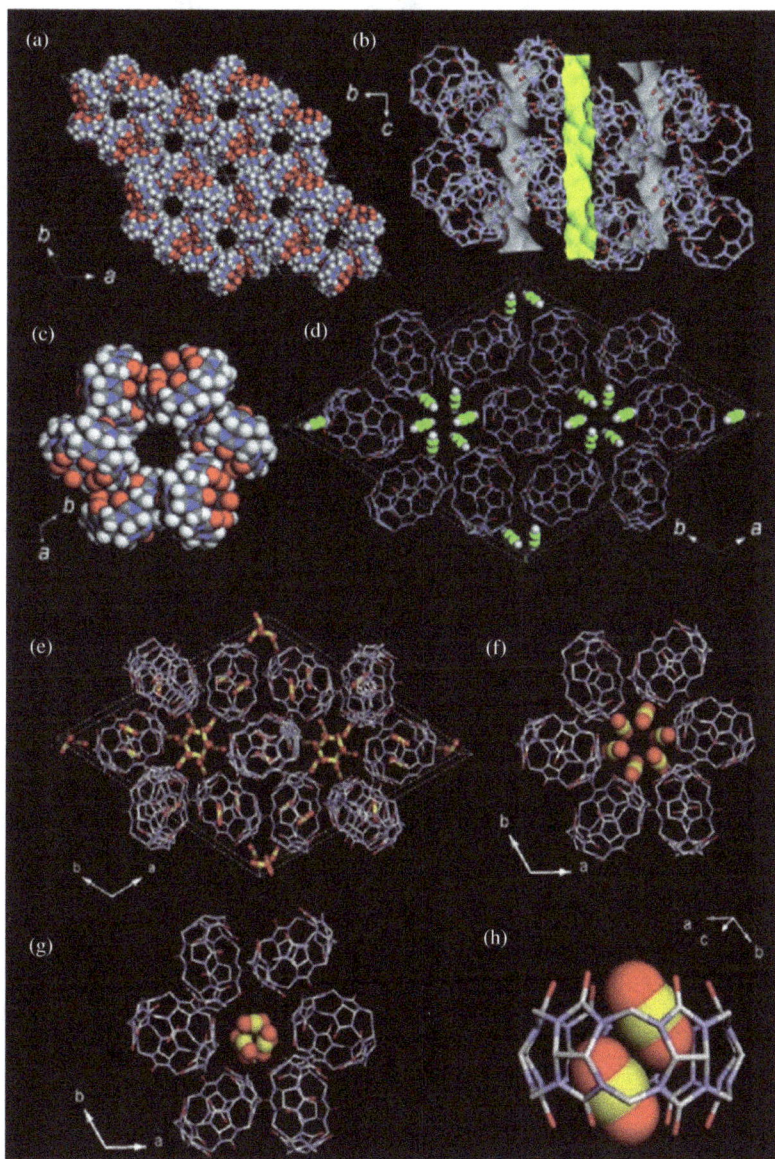

Figure 6.1. (a) X-ray crystal structure of porous CB[6]; (b) 1D channels in porous CB[6] highlighted in yellow; (c) an aperture of a 1D channel which forms a hexagonal arrangement; (d) X-ray crystal structure showing acetylene adsorbed in porous CB[6] (from Ref. 75); (e) X-ray crystal structure of CO_2 adsorbed in CB[6]; (f) sorption site A; (g) sorption site B and (h) sorption site C. (From Ref. 344)

6.2. Cucurbiturils on Planer Surfaces

Immobilisation of cucurbiturils onto surfaces is a powerful way to harness the properties of CB for materials applications. Three strategies have been used to attach CB[*n*] to surfaces (Figure 6.2). The first and most direct is to immobilise the CB[*n*] on a metal (Au and Ag) surface through electrostatic interactions between the polarised ureidyl portal and the metal atoms on the surface. The second way is first to immobilise a guest on a surface and then add the CB[*n*]; this is particularly useful for CB[8] which can form ternary complexes at surfaces. The final strategy is to first modify the equatorial positions of a CB[*n*] and then immobilise through the newly added side chain. In this section, these strategies are described, their merits and drawbacks discussed and the applications that immobilisation facilitates are showcased.

6.2.1. Direct immobilisation

A self-assembled monolayer (SAM) formed by the direct interaction of CB[7] with the metallic surface was first demonstrated by Li *et al.*[76] The monolayer is formed simply by immersion of a gold substrate into a saturated aqueous solution of CB[7]. CB[7] is physisorbed onto the surface through metal–carbonyl interactions, which leads to the formation of a film in which the portal and cavity is accessible. The monolayer was characterised by X-ray photoelectron spectroscopy (XPS) and IR-reflection adsorption spectroscopy. Crucially, a shift in the XPS spectra corresponding to the oxygen 1s peak revealed that the gold–oxygen interaction was

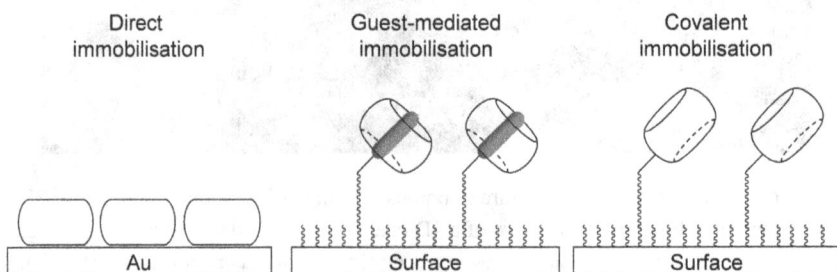

Direct immobilisation	Guest-mediated immobilisation	Covalent immobilisation

Figure 6.2.   The three strategies to immobilise CB[*n*] onto planar surfaces.

the key to the monolayer formation. The monolayer density was estimated to be around 48%, by electrochemically measuring the number of ferrocene molecules that are captured by the surface CB[7]. This figure suggests that CB[7] on gold is an imperfect monolayer. The exact arrangement and density of the CB[7] surface remains a subject of research. Nevertheless, the CB[7] film has been used to immobilise ligands onto a surface, including proteins[347–349] and aptamers[350] for a variety of applications. These studies, which use high-affinity host–guest pairs, are discussed in a later chapter dedicated to high-affinity pairs (Chapter 8).

6.2.2. *Guest-mediated immobilisation*

A strategy that offers more control over the surface coverage is one that first immobilises a guest molecule onto the surface and then adds CB[*n*] from solution. This strategy has the additional advantage of creating stimuli-responsive and reversible surfaces by use of light or redox-responsive guests.

The first example of this strategy was reported by Kim *et al.*[351] who used it to prepare a pseudorotaxane on gold. A spermidine-conjugated 1,2-dithiolane was immobilised onto the gold surface. The thiolane forms a SAM on gold through the Au–S bonds, whereas the spermidine is a high-affinity guest for CB[6]. The addition of CB[6] allowed the formation of the pseudorotaxane since it threads onto the spermidine. The threading was reversed by increasing the pH, which allowed the creation of pH-responsive gate. This gate was used to control the redox chemistry between ions in solution and the gold surface. A similar strategy and ion gating behaviour, this time by immobilising a CB[7] pseudorotaxane onto gold nanorods, has also been reported.[352]

The first example of CB[8] being used to functionalise a surface was reported by Kim *et al.*[310] They grew a polymer perpendicular to the surface using a CB[8]-stabilised charge-transfer (CT) interactions; this work was discussed in Chapter 5. An electrochemically switchable CB[8]-based pseudorotaxanes on gold surfaces has also been demonstrated.[353] A viologen-threaded-CB[8] immobilised on gold could capture electron-rich molecules; reduction of the viologen caused the CT complex to dissociate and thereby release the electron-rich guest. Scherman *et al.*[354] used a related approach to immobilise colloids onto Au substrates in a site-selective manner. In this example, an immobilised methyl violgen was threaded with CB[8], naphthyl-labelled polymeric colloids were then introduced to the surface and anchored to the surface via the surface-bound CB[8]–viologens. Using this method, they could make 1D and 2D patterns on the surface, thereby demonstrating a novel lithographic technique. In a similar manner, Scherman *et al.* also prepared "raspberry-like" colloids on a silica core.[355] The use of azobenzene, methyl viologen and CB[8] as the connectors between the polymeric colloids and the silica core meant that the colloids could be assembled and disassembled by irradiation.

CB[8] has also been used to prepare layer-by-layer protein stacks on surfaces.[356] CB[8] can "glue" together proteins by bridging tryptophan from one protein to a tryptophan residue on a neighbouring protein. They first immobilised a protein via a carboxylate on the protein to an amine surface to form the first layer; they then added CB[8] followed by either the same protein or another protein to create (either homo or hetero) layers of protein glued together by CB[8]. A particularly attractive feature of this work is the lack of protein modification; native proteins can be glued together by CB[8].

Stimuli-responsive surfaces that exploit responsive guests of CB[8] have also been reported. For example, a CB[8]–methyl viologen complex has been used to immobilise proteins[357] through a peptide that controls cell adhesion (Figure 6.3).[92] The use of methyl viologen allows the surface to be controlled electrochemically since the single electron reduction of

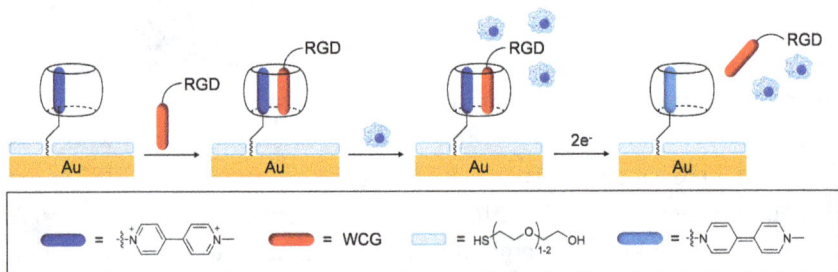

Figure 6.3.   Electrochemically controlled cell adhesion by a redox-active guest molecule in a CB[8] complex.

methyl viologen contained in an immobilised CB[8] complex causes the other guest to be released into the solution. Scherman *et al.* has extended this concept to include light-responsive guests to create orthogonally switchable complexes and surfaces.[333] Azobenzene derivatives, whose alkene bond geometry can switch from *trans* to *cis* upon exposure to UV light, can serve as the π-acceptor in a CB[8]-stabilised CT complex. The *trans* form is a CB[8] guest, but the *cis* form is not. Thus, azobenzenes can be ejected from a CB[8] by *trans*-to-*cis* photoisomerisation. Scherman's group prepared a surface that was functionalised with azobenzene and MV derivatives, and CB[8] could be added from solution. Either light or reductive stimuli caused the complex to disassemble along different pathways. This has allowed the preparation of a reversible, tristable system that may be useful for preparing responsive surfaces at biointerfaces or for rewritable memory devices.

6.2.3. *Covalent immobilisation*

The final approach to attach CB[*n*] to a surface is to tether it covalently through a side chain. The synthesis of functionalised CB[*n*]s can be quite challenging which is why there are fewer examples of this strategy. Various allyloxy-functionalised CB[*n*] have been used for this purpose since they can be efficiently grafted onto various surfaces. The first example of this approach was the immobilisation of perallyoxyCB[6] onto a thiolated glass surface by a thiol–ene reaction.[37] Kim *et al.*[69] also tethered multiallyloxyCB[7] to an allyl-functionalised SAM by employing a metathesis reaction. This surface was then used to immobilise ferrocenylated proteins. The same procedure can also be applied to CB[6] to create a SAM that includes tethered CB[6].[358] Photointiated thiol–ene reactions can also be used to immobilise monoallyoxyCB[7] to a thiol-functionalised polymer.[95] This particular CB[7] surface was used with a corresponding ferrocenylated surface to make adhesive surfaces, termed supramolecular Velcro. The supramolecular Velcro and protein immobilisation examples use high-affinity host–guest pairs and are discussed in more detail in a later chapter (Chapter 8).

Unfunctionalised CB[7] has been covalently tethered to a surface by a photochemical reaction.[359] An aqueous solution of CB[7] was incubated on an azido-functionalised surface under irradiation with UV light. The presence of CB[7] on the surface was confirmed by treatment with acridine orange, a fluorescent guest for CB[7]. While there was no direct evidence for the mechanism of immobilisation, a reasonable assumption is that UV irradiation of the azide yielded a highly reactive nitrene which then inserts into one of the methine CH-bonds on the CB[7]. This work represents a facile method of immobilising CB[7] onto a surface without the need for a functionalised CB[7]. Since the reaction is light-mediated, photolithographic patterning methods can be used to fabricate devices.

6.2.4. *Measurement of host–guest interactions on surfaces*

For most of the applications of CB[n] on surfaces, it is assumed that the immobilised CB retains its binding affinity and host–guest behaviour. However, this may not always be the case, particularly with the CB[n] immobilised though carbonyl–metal interactions. These CB[n] may have reduced affinity since the "bottom" portal is less polarised and larger guests that would protrude out of the bottom may not have the room to do this when the CB[n] is on the surface.

Sophisticated surface characterisation techniques have been employed to measure the affinity of host–guest interactions on a surface and to examine the topology of CB[n] surfaces. Atomic force microscopy was used to measure the rupture force between CB[6] and spermine.[358] The gold AFM tip was functionalised with 1,2-dithiolane spermine and the CB[6] was covalently tethered to the surface. The rupture force required to separate a single complex was measured to be 120 pN which at the time was the highest measured for a synthetic host–guest system on a surface. A similar technique has been applied to CB[7] SAMs.[360] The CB[7], in this case, was directly immobilised on the gold through carbonyl–metal interactions and the tip was functionalised with an adamantane. The rupture force was measured to be 140 pN (at 45 000 pN s^{-1}) and because the CB[7] monolayer is not well packed (~48% from the study cited above and 40 ± 9% from this study) there was an additional interaction measured between the gold surface and the adamantane tip. The dissociation rates of the Au–adamantane and CB[7]–adamantane were calculated to be

0.3 s^{-1} and 0.03 s^{-1}, respectively. The difference in these values suggests that the strength and stability of the CB[7]–Ad interaction is sufficiently stronger than the non-specific interactions, so the low surface coverage is not a concern. The spontaneous absorption of CB[6] and CB[7] has been compared by AFM.[361] A monolayer was achieved with a saturated aqueous solution of CB[6]. The monolayer could be used for sensing although the defect sites had to be filled with an additive. The much more soluble CB[7] formed a multilayer on the surface, when left for the same amount of time as CB[6]. However, a homogeneous monolayer suitable for sensing could be prepared by a shorter incubation time. The same group have examined CB[8] as a sensor surface by AFM and electrochemical methods.[362] They showed that CB[8] can form a sensor surface on various substrates. Recently, the binding and electrochemical behaviour of CB[7]–Fc complexes at an electrode interface was investigated.[363] It was found that the rate of association of the complex is slower at an interface than in solution, whereas there is no change in the dissociation rate. A SAM that presented Fc moieties was used as a platform to measure binding at the surface. After the surface was treated with CB[7], voltammetric measurements were performed on the Fc@CB[7], which revealed near-ideal redox behaviour. Importantly, the covalently tethered complex was found to be of higher affinity than that of a physio-adsorbed complex, suggesting that this is the superior strategy to harness the properties of CB[n] on a surface.

In summary, the direct interactions between CB[n] portals and gold is a phenomenon that has been exploited in the formation of CB[n] SAMs. While it shows promise as a strategy to immobilise molecules of interest, fundamental studies still need to be done. More advanced surface characterisation techniques may facilitate more detailed investigation into the properties and topology of CB[n] surfaces. However, a limited number of AFM and electrochemical studies seem to show that CB[n] retain their host–guest chemistry reasonably well when immobilised on a surface.

To immobilise a CB[n] and keep the original properties as far as possible, the best approach appears to be covalent tethering through a side chain. This approach has led to many interesting applications and can harness the unique properties of CB[n] for materials and biological science applications. A drawback to this method is the need to functionalise the CB[*n*] of interest with chemical handles such a hydroxyl or an allyloxy group.

It is worth noting that CB[8]'s ability to form high-affinity ternary complexes is rare among synthetic host–guest systems. CB[8] can facilitate the non-covalent coupling of an appropriately functionalised surface to another species to prepare functionalised surfaces or composite materials. The other feature that can be exploited is CB[8]'s affinity for redox-responsive (methyl viologen) and photo-responsive guests (azobenzene), which can be exploited to prepare stimuli-responsive interfaces.

6.3. Cucurbiturils on Metallic Nanoparticles

In the previous section, it was shown that CB[*n*] can be immobilised on metallic surface through carbonyl–metal interactions. The same interactions can be used to immobilise CB[*n*] onto metallic nanoparticles, or even bridge together metallic nanoparticles to form clusters. The formed assemblies have interesting applications in catalysis and photonics.

6.3.1. *CB–metal nanoparticle assemblies*

Mahajan *et al.*[76] have shown that CB[*n*] (*n* = 6–8) can act as a "glue" between adjacent nanoparticles resulting in small gold nanoparticle clusters. The formation of these clusters was clearly seen by TEM imaging. CB[*n*] have also been used as a capping agent and stabilising ligand in the formation of nanoparticles. For example, CuO nanoparticles (5 nm diameter) can be formed with the assistance of CB[7] as bridging group.[364] CB[7] and Cu^{2+} acetate first form a complex in a 2:1 ratio as a precursor to the nanoparticle formation at high temperature. Using a similar procedure, silver[365] and gold nanoparticles have also been prepared.[366] These procedures resulted in well-dispersed nanoparticles with well-defined sizes, although mechanistic details have yet to be ascertained.

Aggregates of gold nanoparticles bridged by CB[5] have been prepared (Figure 6.4). CB[5] acted as a capping agent, and the size of the nanoparticles could be controlled by adjusting the ratio of CB[5] and the gold salt. The aggregation process was monitored by measuring the

CB[*n*]-induced gold nanoparticle cluster formation

(a)

(b)   (c)

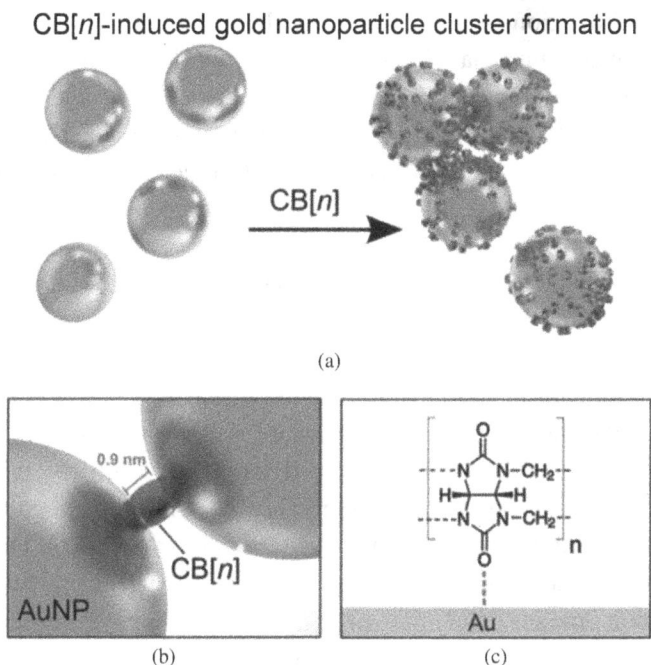

Figure 6.4.   (a) CB[*n*]-bridged nanoparticles. (b) A consistent (0.9 nm) gap between adjacent nanoparticles. (c) Portal-mediated Au-CB[*n*] interactions. (Modified from Ref. 367)

surface plasmon resonance (SPR) of the colloidal solution. Monodisperse gold nanoparticles absorb in the green region, whereas the aggregated forms undergo a red shift. This can be seen in an SPR sensorgram and also visually since the solutions change colour from red to purple/blue. The mechanism of this process was investigated in detail.[85] The main conclusions were that the reaction kinetics were controlled by the concentration of CB[5], and the rigid CB[5] ensured that the distances between the gold particles were consistent, which resulted in distinct plasmon resonances. In more recent work, the same procedure was extended to CB[*n*] (*n* = 6–8).[368] In this work, the inter-particle distances were measured to be ~1 nm in all cases, which accords well with the height of CB[*n*] (0.9 nm) and provides further evidence that the portal–metal binding is the key interaction in the formation of the aggregates.

The stabilisation of silver nanoparticles by CB[n] has been investigated.[369] It was found that CB[5] and CB[6] poorly stabilise silver nanoparticles, but the more flexible family members, CB[7] and CB[8], can undergo minor structural distortions to accommodate better the metallic surface. Surprisingly, the presence of a guest in the CB[7] cavity has little effect on its ability to stabilise nanoparticles; even if a guest is present, some of the carbonyls are still free to interact with the metal. Computational studies support the idea that the interactions between the CB[7] carbonyls and the silver atoms are the key to the assemblies' stability.

CB[n] nanoparticle assemblies exhibit several interesting properties. The nanogaps of such assemblies have interesting photonic properties. The electric field in the nanogaps between these particles is strongly enhanced.[370] The well-defined length (0.9 nm) of the CB[n] spacer ensures that the local field enhancement in the nanogap is consistent, while potentially allowing space for small molecules. Therefore, CB[n] provide a consistent nanogap that facilitates the investigation of plasmon modes and the near-field enhancement properties.[371] Furthermore, CB[n] can template the threading together of gold nanoparticles to yield a continuous, conducting gold string.[372] Individual nanoparticles were arranged into a chain by addition of CB[7]; they were then "welded" together by irradiation with a femtosecond laser (Figure 6.5(a)). The irradiation caused the neighbouring nanoparticles to merge because of the enhanced field effects between the particles. The 0.9 nm gap between the particles can clearly be seen in TEM images before irritation; after irradiation, the particles are a single entity (Figure 6.5(b)). Interestingly, the width of the strings can be controlled by the wavelength of the laser. Furthermore, the field enhancement effects in the cavities and crevices of these structures suggest that they may be used in photovoltaic devices.

Self-assembled monolayers (SAMs) can also be formed on metallic particles. SAMs functionalised with CB[n] guest can, therefore, be used to decorate nanoparticles with CB[n]. Rotello *et al.*[373] have used nanoparticles functionalised with alkylhexadiamine to mediate the interactions of nanoparticles with proteins. CB[7]-decorated nanoparticles interact with proteins with higher affinity and higher stoichiometry than those without CB[7].

(a)

(b)

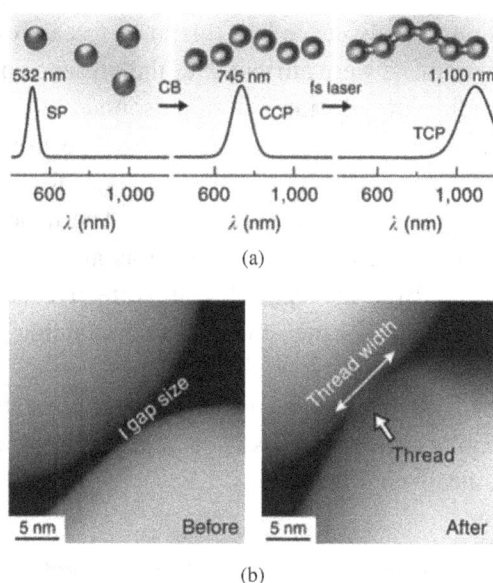

Figure 6.5.   (a) Surface plasmon resonances of Au nanoparticles, CB[n]-linked Au nano-particles, and a Au nanostring. (b) TEM images of CB[n]-linked nanoparticles and Au nanostring, before and after femtosecond laser irradiation, respectively. (From Ref. 372)

Gold nanorods functionalised on the ends with methyl viologen could be linked together by binaphthyl units and CB[8].[374] The three-component system was capable of aligning nanorods reversibly; once a competitive guest was added, the assemblies dissociated. Furthermore, the rigidity of the assemblies could be controlled by adjusting the length of the dinaphthyl linker.

6.3.2. Applications in catalysis

Such CB-capped nanoparticles appear to have applications in catalysis. For example, Pd nanoparticles capped with CB[6] have been used to pro-mote Suzuki–Miyaura cross-coupling reactions.[375] The CB[6] seems to protect the active catalyst from aggregation while allowing reactants access the metal surface; furthermore, the assemblies are recyclable. Platinum-based nanoparticles have also been prepared with CB[6] as the stabilising group. These assemblies not only showed catalytic activity for methanol oxidation, but also excellent tolerance to catalyst poisoning.[376] Both of these phenomena were attributed to the presence of CB[6] which promoted the binding of substrates over poisonous intermediates. Gold nanoparticles capped and bridged by CB[7] and free from other organic ligands or metal cations appear to be better catalysts for reduction of dissolved oxygen than the nanoparticles alone.[377] The nanoparticle–CB[7] assemblies encapsulate O_2 and thereby serve as an O_2 delivery vehicle. Palladium and Au–CB[7] nanoparticles are effective catalysts for the reduction of nitrophenol[378] and other nitro-containing compounds,[379] respectively.

Decamethyl CB[5] has also been used to prepare catalytic nanoparticles. Such assemblies have been used to catalyse Heck cross-coupling reactions[380] and nitrophenol reductions.[381]

6.3.3. *Applications to surface-enhanced Raman spectroscopy*

Raman spectroscopy is a vibrational spectroscopy technique that, unlike infrared spectroscopy, is not affected by water molecules in the fingerprint region. However, a limitation of this technique is its low sensitivity. One way to enhance the sensitivity is to measure the analyte near a plasmonic surface, such as a metallic surface or nanoparticle. The resulting plasmonic field enhancement can be as high as 10^{11},[382] which is enough to enable single molecule sensing for most analytes.[383] For such a plasmonic enhancement, the molecule of interest should be localised close to the plasmonic surface.

In the previous section, the plasmonic field enhancement effect in the 0.9 nm nanogaps between CB[n]-separated nanoparticles was described. In this section, the exploitation of these effects for surface-enhanced Raman spectroscopy (SERS) is discussed. The nanogap where plasmonic effects are manifested is termed a "hot spot", and CB[n] can serve a dual role (Figure 6.6). Firstly, it creates a well-defined nanogap of 0.9 nm; secondly, it serves to localise guest molecules inside the hot spot to facilitate single molecule sensing. Scherman et al.[384] provided the first report of SERS being combined with CB[n]. They observed clearly identifiable peaks corresponding to the CB[n]'s vibration mode which allows the individual CB[n] to be assigned. A SERS spectrum for a guest encapsulated

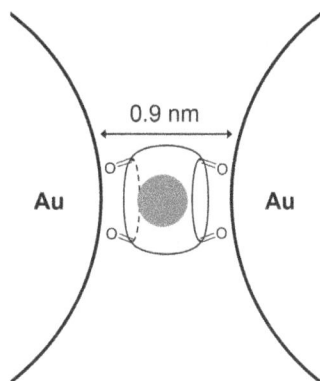

Figure 6.6. The formation of a plasmonic hot spot in the gap between metallic species using CB[n] as a molecular spacer.

in CB[7] was first demonstrated by Mahajan *et al.*[85] CB[7]-capped gold nanoparticles were treated with Rhodamine 6G (a CB[7] guest), which localised the guest in the SERS hot spot and significantly enhanced the Raman spectrum of the guest. Another example utilised neutral ferrocene in CB[7] and then used the complex as a bridge between nanoparticles to create a hot spot.[385] A 10^9-fold enhancement was observed in the Raman spectrum of ferrocene. Without CB[7] to localise the ferrocene, a SERS spectrum could not be obtained because ferrocene has little affinity for metallic surfaces.

SERS-sensing is an incredibly sensitive method for studying host–guest interactions. Since SERS measures the vibrational modes of the CB[n], the increase in bond strain upon guest encapsulation can be seen in the SERS spectrum. Such effects have been predicted computation-ally,[386] and have been observed experimentally with CB[8],[387] albeit not strongly in the solid state. The shifts in the Raman peaks are more appar-ent in the guest molecules, than in the host itself.[388] The shift upon host–guest complexation can be used to quantify the extent of guest encapsulation. For example, with polymathic hydrocarbon guests. SERS has also been used to measure the encapsulation of diquat (a pesticide) into CB[8]-capped silver nanoparticles.[389] Furthermore, the pH depend-ence of this guest encapsulation was examined using the SERS method.

Photochemical reactions have also been examined within a plasmonic nanojunction using SERS.[390] Recently, strong coupling was observed between an emissive dye held in a CB nanocavity and the light field, which allowed quantum effects to be manifested at room temperature.[391]

The combination of SERS and CB[*n*] is a potent one that allows accurate, multiplexed analysis of single molecules. In the future, such a combination may be used to investigate dynamic processes at the single or few molecule level.

6.4. Polymeric Nanomaterials

Self-assembled nanostructures are often formed using weak, reversible interactions; this allows for error correction and self-checking mechanisms.[392,393] However, these weak interactions rarely lead to robust nanomaterials that are useful for practical applications. An emerging approach which has been pioneered by Kim *et al.*[394] is the formation of well-defined nanostructured materials using covalent bonds. Despite the lack of an error-correcting mechanism, the use of rigid building blocks, such as CB[6], with a rigid linker has resulted in robust materials, such as capsules and polymer films. The cavities of the CB units are still free and accessible, which means that the nanostructures can be non-covalently modified. Supramolecular architectures that are assembled at the interface of microdroplets in microfluidic devices are also considered in this section since they represent a powerful combination of microfluidics and host–guest chemistry.

6.4.1. *CB[6]-based polymer nanocapsules*

The first example of self-assembly using covalent chemistry was the preparation of spherical polymer nanocapsules from functionalised cucurbiturils (Figure 6.7).[71] CB[6] functionalised around its periphery leads to the formation of a disc-shaped molecule with in-plane functional groups. Allyloxy-functionalised CBs are particularly useful for this chemistry since they are substrates for the thiol–ene reaction. Because the CBs are functionalised through the equatorial positions on the ring, the cavity remains available for non-covalent modification, which allows for the modification of the nanostructure surface properties.

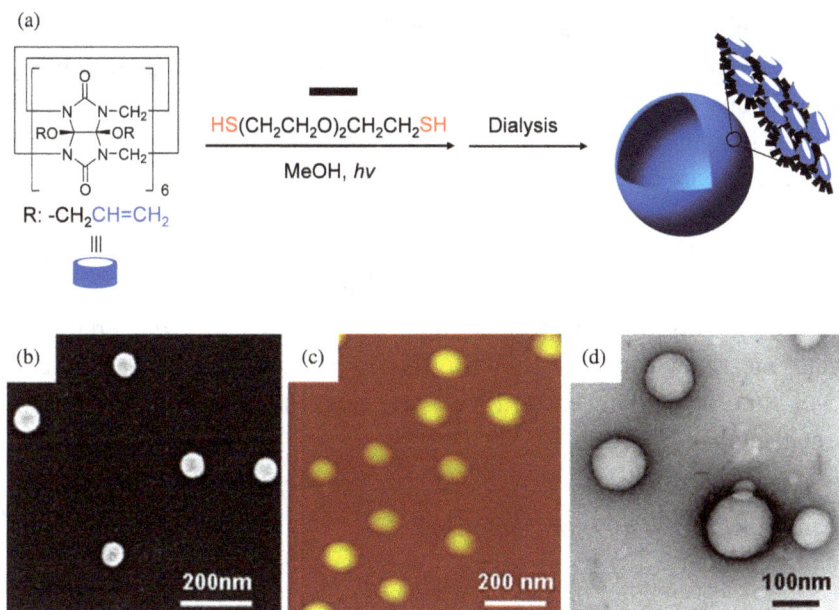

Figure 6.7. (a) Preparation of CB-based polymer nanocapsule; (b) SEM; (c) AFM and (d) TEM images of the nanocapsules. (From Ref. 71)

PerallyoxyCB[6] and a dithiol linker spontaneously form hollow nanocapsules, without a template, when irradiated with UV light.[71] The nanocapsules are formed in methanol in excellent yield (>85%) and have quite a narrow size distribution (110 ± 30 nm). The size of the nanocapsules can be easily tuned by changing the solvent or the length of the dithiol linker. For example, the nanocapsules are much larger (500 nm) in a better solvent, such as $CHCl_3$, or can be made smaller by shortening the length of the dithiol linker (50 ± 10 nm for ethanedithiol).

Experimental and theoretical studies have provided some insights into the formation mechanism. SEM and DLS revealed that the key intermediates are slightly curved oligomeric patches, which then assemble into spherical nanocapsules.[71,395] During the early stages of assembly, the CB[6] discs form dimers and trimers, which then grow in lateral dimensions into 2D patches. As these patches continue to grow, they begin to curve; further covalent crosslinking of these patches generates the final spherical nanocapsule.

The nanocapsule is a water-dispersible robust nanostructure, the surface of which can be non-covalently modified. As such, there are many applications in biology; for example, the surface can be modified with various imaging and targeting groups for multimodal bioimaging.[396] Such applications are discussed in detail in the next chapter.

The nanocapsules can also be covalently modified through "loose-end" thiols. The thiol groups can be used to decorate the nanocapsule surface with metallic nanoparticles, including Pd, Au and Pt. Pd-functionalised nanocapsules, in particular, are stable, dispersible in water and show excellent catalytic activity and reusability for C–C and C–N bond formation reactions in water, suggesting their potential as a green catalyst.[397]

6.4.2. *CB[6]-based 2D polymer films*

Free-standing, single-monomer-thick, 2D polymers with a covalently linked 2D network structure have attracted considerable attention in recent years,[398] because of their potential in molecular electronics, selective transport materials, sensors and surface catalysts.

In the previous section, the solvent dependency on the formation of the size of nanocapsules was shown. In the early stage of the self-assembly, a 2D-patch is formed, which raises the question: Could these patches be grown laterally into films by changing the reaction conditions or reaction components? Early work using DMF as the solvent resulted in what appeared to be rolled or folded scroll-like structures.[394] With these thoughts in mind, a 2D, free-standing, CB-containing polymer was prepared (Figure 6.8).[94] To construct these polymers, several design principles were employed: (i) rigid disc-shaped building blocks with in-plane reactive groups, in a single plane; (ii) short linkers to prevent curvature; and (iii) a good solvent that would allow the growing 2D-patches to remain flat and grow in lateral dimensions. Indeed, when perallyloxyCB[6] was reacted with 1,2-dithioethane in DMA, under UV irradiation, a 2D polymer that was tens of micrometer across was formed.

The 2D polymer was analysed by AFM and TEM. Initially, AFM showed that the polymers formed were more than a single-layer thick, presumably because of multilayer stacking. To make a single-layer thick polymer, spermine guests were incorporated to add a positive charge on

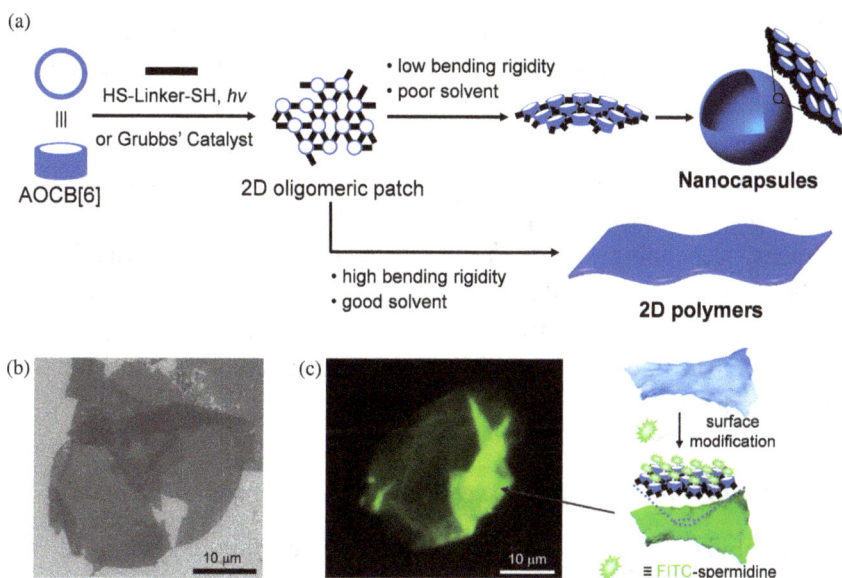

Figure 6.8. (a) Preparation of CB-based, 2D polymers, (b) SEM and (c) fluorescence microscope image of FITC-spermidine labelled polymer. (Modified from Ref. 394)

each side of the film, thereby causing electrostatic repulsion between the films. When measured by AFM, the height of the film was measured to be 2.0 nm, which is in reasonable agreement with the predicted height of a single monomer of the guest-complexed polymer. The internal structure was examined by high-angle annular dark-field scanning-TEM (HAADF-TEM) after the CB[6]s were labelled with gold nanoparticles. This imaging method revealed that, despite many defect sites, the nanoparticles were distributed evenly with approximately six neighbours each, separated by 3 nm, which is in good agreement with the calculated hexagonal arrangement of the repeating unit.

The 2D polymers are useful as permselective membranes.[399] The hexagonal network structure that is formed has two different pores, the CB[6] cavity and the gap between adjacent CB[6] units, which engenders size selectivity to membranes based on these polymers. Moreover, the CB[6] cavity can be used to functionalise the membrane with various functionalities including positively and negatively charged groups. This functionalisation allows the membrane to be tailored to enable the selective permeation of dye molecules based on charge and hydrophobicity. Another use of these polymers is the absorption of dye molecules in water remediation applications.[282]

When the linker molecules used has a reversible bond, nanostructures with dynamic morphology are obtained.[104] Thiol-functionalised CB[6] can form nanocapsules or polymer sheets spontaneously through disulphide linkages, depending on the reaction medium. Most importantly, by exchanging the solvent, the nanostructure morphology can be switched from film to capsule or *vice versa* by virtue of it being prepared by dynamic covalent bonds, which allows reorganisation of the nanostructure to the thermodynamic minimum.

6.4.3. *CB[8]-microcapsules assembled within microfluidic droplets*

The combination of non-miscible liquids in microfluidic systems allows the formation of microdroplets, into which cargo can be enclosed. Scherman and Abell have pioneered the integration of such microdroplets with interfacial host–guest chemistry to prepare various supramolecular architectures (see Ref. 400 for more details).

The interfacial synthesis of polymer–nanoparticle composite materials has been reported.[88] In a microfluidic system, the mixing of water and a fluorous oil results in the formation of mono-dispersed water droplets. Aqueous solutions of CB[8], viologen-functionalised nanoparticles, and a naphthyl-functionalised polymer were mixed with a fluorous oil which resulted in the formation of microcapsules around the droplets. Dehydration afforded stable hollow microcapsules, albeit with a reduced size (Figure 6.9(a)). Various molecular cargos could be encapsulated in the microcapsules by the addition of an aqueous stream to the microfluidic device that contained a cargo molecule. Because of the CB[8] complex, the capsules could be degraded by external stimuli. This work represents a facile, one-step method of preparing micrometer-sized supramolecular architectures.

Work has continued along these lines by changing the nanoparticle component to another polymer, which allowed much greater diversity of microcapsules since polymers are easily synthetically modified (Figure 6.9(b)).[401] The mixed polymer microcapsules were prepared by

Polymer-nanoparticle microcapsules

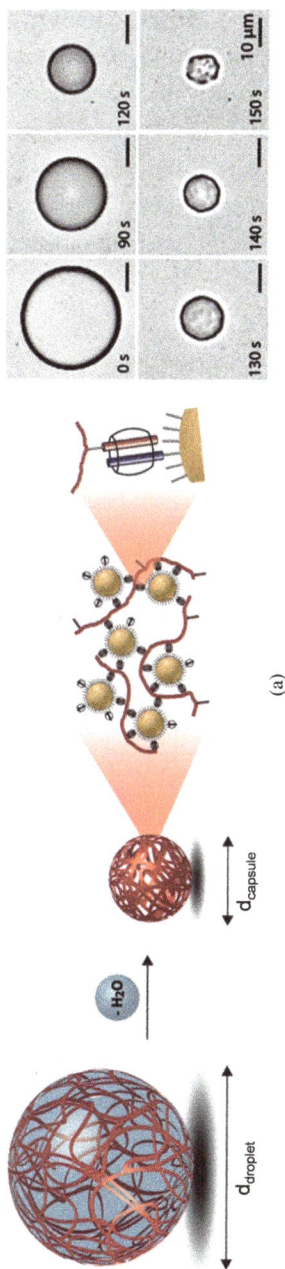

(a)

Layered assembly polymer microcapsules

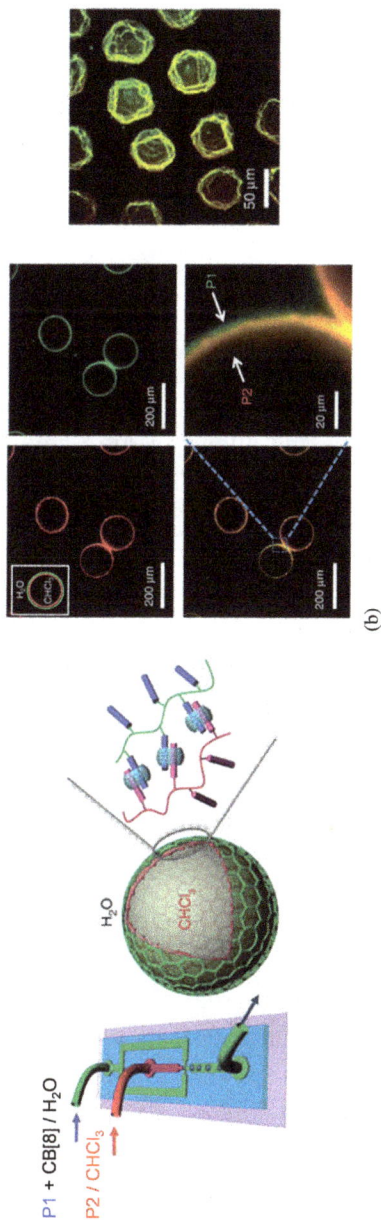

(b)

Figure 6.9. (a) The preparation of microdroplet templated polymer-nanoparticle microcapsules prepared in a microfluidic device; dehydration of the droplet resulted in hollow capsules. (From Ref. 88) (b) Layered assembly of polymer–polymer microcapsules prepared in a mixed solvent microfluidic device. (From Ref. 401)

mixing organic-soluble and water-soluble components; varying the solvent ratio afforded either water droplets or organic droplets, which meant that either water-soluble or organic-soluble cargos could be contained inside the capsule. The combination of mixed-phase microfluidics and interfacial microfluidics is powerful and allows the construction of stimuli-responsive microscale capsules. The main limitation is in the microfluidic technology: currently, microscale materials can be prepared routinely, but nanoscale is much more challenging without specialist equipment capable of preparing nanodroplets.

6.5. Hydrogel Materials

Hydrogels are made from crosslinked hydrophilic polymers and contain a substantial amount of water. The high water content makes them biocompatible and allows them to be used in biomedical applications for drug/gene delivery, cell therapy and tissue engineering. Typically, they are formed by covalent crosslinking which makes them quite robust, but their chemical and physical properties are difficult to tune. Supramolecular hydrogels formed by non-covalent crosslinking including hydrogen bonding, hydrophobic interactions and host–guest interactions have emerged as a novel soft biomaterial showing more dynamic properties. CB[6], CB[7], CB[8] and *nor-seco*-CB[10] (*ns*-CB[10]) have all been used as supramolecular crosslinkers in the perpetration of new soft materials for 3D cell culture, cell delivery, tissue engineering and even wood preservatives. This section details these hydrogels and their applications.

6.5.1. *CB[6]-containing hydrogels*

CB[6] has been used to modulate the properties of a small molecule gelator.[402] Pseudorotaxanes formed by butan-1-aminium-4-methylbenzenesulphonate (BAMB) threaded into CB[6] (BAMB@CB[6]) work as a small hydrogelator which self-assembles into fibrils in water by combining multiple non-covalent interactions such as hydrophobic interactions and π–π stacking among BAMB, van der Waals interactions between BAMB@CB[6] and hydrogen bonding between water and CB[6].

Another strategy that offers more control of the hydrogel properties is to crosslink host-functionalised polymers with guest-functionalised polymers. Another advantage of this strategy is that uncrosslinked guest molecules that remained in the hydrogel can be used to modify the chemical properties of the hydrogel by treating functional-tag-conjugated host molecules. Kim *et al.*[35] described how AOCB[6] could be grafted to thiolated hyaluronic acid (a widely used polysaccharide biopolymer for hydrogel formation) by a thiol–ene photoreaction to yield a CB[6]-anchored linear polymer with an available CB[6] cavity (HA–CB[6]) (Figure 6.10). A complimentary diaminohexane (DAH)-modified polymer was also prepared (HA–DAH). When these two polymers were mixed, gelation occurred within two minutes. The host–guest interaction was crucial for gelation, which was shown by a gel–sol transition that occurred when an excess of a competitor guest (spermidine) was added. The hydrogel was mechanically robust enough to hold its shape without any scaffold. Furthermore, it can entrap live cells inside on forming a hydrogel and control their spreading behaviour upon treatment with a functional-tag-conjugated CB[6] such as cyclic RGDyK-conjugated CB[6], which promotes cell adhesion. Furthermore, none of these components causes significant cytotoxicity. Hydrogel formation was also tested in a live animal, and the hydrogel retained its shape for almost two weeks and did not appear to be toxic to the animal. Taken together, these results demonstrate that this self-assembling hydrogel is a good soft biomaterial for 3D cell entrapment and culture.[35] Encouraged by this result, the CB[6]–HA

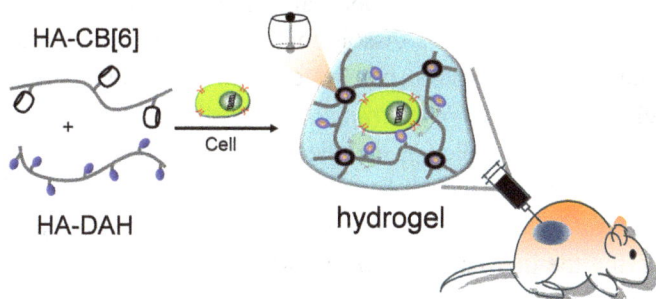

Figure 6.10. CB[6]-containing hydrogels that are robust enough to form and persist in live animals.

hydrogel was used for biomedical applications such as tissue engineering and cell therapy in live mice.

When human mesenchymal stem cells (hMSC) were entrapped in the self-assembling hydrogel, their differentiation to chondrogenesis was controlled by treatment with drug-conjugated CB[6].[403] In addition, when engineered MSCs (eMSCs), which release the cancer-treating protein, interleukin-12 (IL-12), were entrapped in the hydrogel, the cells produced IL-12 for more than 40 days, resulting in inhibition of tumour growth and significantly enhanced survival rate in cancer-bearing mice.[404,405] These experiments demonstrated the feasibility of CB[6]-based hydrogels as emerging 3D cell culture scaffold for stem cell therapies and tissue engineering.

6.5.2. *CB[7]-containing hydrogels*

The first CB[n]-containing hydrogel of any kind was reported by Kim *et al.* in 2007.[70] They used CB[7] as a small gelator (3–5 wt.%) to form a hydrogel in a warm, dilute solution of mineral acid (such as sulphuric acid (0.5 N)) upon cooling. No other member of the CB family can do this in any solvent. It was found that CB[7] self-assembled into a fibre, then formed a bundle of fibrils which hold a significant amount of water to form a hydrogel. It can be attributed to the non-covalent interactions of CB[7] with neighbouring CB[7] directly via C–H··O hydrogen bonding and indirectly through water/hydronium ions as revealed by the X-ray crystal structure. The formation of this hydrogel can be controlled in several ways. Firstly, it is pH-sensitive, and its optimal formation is pH 2. The presence of alkali metal ions promotes dissolution of the hydrogel. More interestingly, the addition of a photoresponsive guest, 4-diaminostilbene, can control the gel–sol transition. When *trans*-diaminiostilbene, which is known to form a 1:1 inclusion complex, is included, a white gel is obtained, but when it is photoisomerised to the *cis* form under UV light, a yellow solution is yielded from the gel; the process can be reversed by heating the mixture and promoting the thermal *cis*-to-*trans* isomerisation. In all these cases, it is disruption of the H-bonding network that leads to the gel–sol transition.

Appropriately functionalised polymers can be non-covalently crosslinked by using the strong host–guest interactions between CB[7] and aminoadamantane.[406] However, fast kinetics of the non-covalent interactions between two polymers (one with CB[7] conjugated and the other with aminoadamantane conjugated) did not form well-crosslinked hydrogels but instead provided aggregates. Pretreatment of the CB[7] with interfering guests such as methyl viologen, DAH and ferrocene derivatives slowed down the hydrogel formation and eventually allowed the formation of a well-crosslinked hydrogel; the mechanical properties could be tuned by the amount and strength of the interfering guest. The slow hydrogel formation allowed interesting applications such as printed hydrogels that gelated upon standing.

6.5.3. CB[8]-containing hydrogels

The remarkable host–guest chemistry of CB[8], which can accommodate two different molecules such as methyl viologen (MV) and naphthalene (Np), presents another opportunity to form supramolecular hydrogels. Scherman *et al.*[34] first demonstrated the formation of polymeric hydrogels from two differentially functionalised polymers with CB[8] as a non-covalent crosslinker to "handcuff" the polymers together (Figure 6.11). One polymer was functionalised with MV and the other with Np. They observed the formation of a hydrogel from an aqueous solution containing

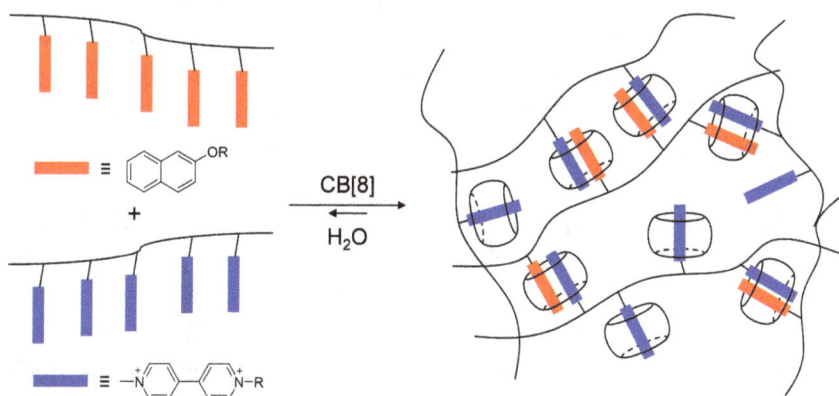

Figure. 6.11.   Hydrogel formation mediated by host–guest complexation with CB[8]. (From Ref. 34)

the modified polymers, only after CB[8] (5 wt.%) was added, demonstrating the importance of CB[8] acting as a crosslinker for the gelation. Further investigation yielded ultrahigh water content (99.7%) hydrogel prepared from cellulose polymers.[407] In addition, the hydrogel's mechanical properties could be tuned over three orders of magnitudes by varying the loading of three gelation components. Furthermore, they have excellent shear thinning properties due to their low polymer loading, and they also exhibit self-healing behaviour by recovering their host–guest interactions efficiently. The hydrogels thus formed are responsive to several external stimuli such as temperature, redox chemistry, and competing guests. This collection of properties suggested that these hydrogels may be useful for biomedical applications such as sustained release of proteins, in analogy with the CB[6]-based system, described earlier.

CB[8]-mediated network formation can be combined with another non-covalent modality to prepare a double-networked hydrogel; namely, a DNA-hybridised system.[408] The double network exhibits multiple levels of stimuli responsiveness including temperature, chemical and enzymatic responses. Although interpenetrated, both networks existed orthogonally and did not interfere with each other. The mechanical strength of the formed material was greater than that predicted based on the sum of individual networks, which is one of the characteristic properties of double-networked soft materials. Furthermore, the gel had the ability to reorganise and self-heal rapidly after sustaining damage. External stimuli could modulate the properties of the gel: addition of excess phenylalanine disrupted the CB[8] network, and DNAase treatment digested the DNA network. The authors suggest that this soft material may be another scaffold for 3D cell culture, and may be used as an injectable delivery device for DNA-based therapies.

The utility of CB[8] hydrogels is not limited to biomedical applications. A CB[8]-based hydrogel has also been used as a preservative for wood.[409] The hydrogel was prepared from an antibacterial polymer since bacteria cause wood degradation. The first guest was MV, and there were two second guests for CB[8], Np- and catechol-based guests which dynamically competed with each other for CB[8] binding. However, in the presence of Fe^{3+}, another wood degradant, the catechol sequesters Fe^{3+} (since catechol acts as a ligand for Fe^{3+}) and stops it spreading through the wood, which also strengthens the hydrogel by creating another interaction network. Another consideration is how to keep the wood wet, since drying causes damage. Normally this is achieved by the hydrophilic polymer, PEG. The high water content hydrogel works in the same manner. They demonstrated their material on a real wooden artefact: wood from a 16th century warship. The hydrogel has clear advantages over currently used PEG-based preservatives since it also sequesters iron and has anti-bacterial properties.

These hydrogels are non-covalently crosslinked dynamic soft materials, and their dynamics influence their physical properties. There is a relationship between binding kinetics and their macroscopic physical properties: A slow rate of dissociation leads to increased mechanical strength, and a fast rate of association is beneficial for efficient self-healing materials.[410] A follow-up investigation focused on the importance of crosslinking dynamics on cargo release from a hydrogel.[411] As expected, it was easier for a drug molecule to diffuse through hydrogels with lower dissociation energy. Furthermore, they showed that hydrogels with low dissociation rates erode faster. These findings indicate that the dynamics of the crosslinking is an important consideration and can be used to tailor the properties of the hydrogel.

A photo-activated hydrogel was constructed using anthracene-derivatised cellulose polymer.[412] In the presence of CB[8], an elastic hydrogel forms spontaneously. After exposure with UV light the material "hardened" as the anthracenes covalently dimerised inside the CB[8] cavity, resulting in three-fold increase in the storage moduli. Since the modification was covalent, transition was irreversible. While a hydrogel can be formed by irradiation in the absence of CB[8], CB[8] accelerates the reaction.

Very recently, a very tough and self-healing CB[8]-hydrogel was reported.[413] A biomimetic approach was used to prepare a dual network hydrogel, one covalently linked and the other dynamic and non-covalent. When force is applied to this material, the weak non-covalent bonds break, whereas the covalent bonds do not, which allows the material to stretch; when the material is relaxed, the non-covalent bonds reform and the material recovers its original properties. Later, the mechanical and conductivity properties of a similar hydrogel was examined. It was found that such hydrogels could be assembled *in situ* with as little as 2.5% CB[8] to give self-healing materials with remarkable toughness and stretchability.[414]

6.5.4. *Other hydrogels*

Another type of supramolecular hydrogel consists of four-arm adamataneamine-functionalised polymers crosslinked with CB[10] (*ns*-CB[10]).[415] In this case, *ns*-CB[10] acts as a non-covalent crosslinker, similar to CB[8], but it can contain a greater variety of guests than CB[8] because of its larger cavity. This hydrogel also exhibited efficient self-healing.

In summary, CB[6]-based hydrogels have led the way in the biomedical applications. A simple cell-compatible polymer can be functionalised with CB[6], and a polyamine; simply mixing these polymers yields a hydrogel that can be used as a 3D cell culture scaffold. The hydrogel can be fine-tuned by post-synthetic modification with various CB[6]-tags. Such efforts have led to the demonstration of hydrogels being used as a scaffold for stem cell therapies, even in live animals. CB[8] had shown great scope to control the properties of hydrogels utilising dynamic, reversible and controllable binding behaviour among the three gelation components. Several different scaffolds have been functionalised with various guests, leading to materials with various stimuli responsiveness and mechanical properties. The future of CB[*n*]-based hydrogels seems bright and real biomedical applications are anticipated.

Chapter 7

Biological Applications of Cucurbiturils

7.1. Recognition of Amino Acids, Peptides and Proteins by CB[*n*] and Applications

Amino acids are the building blocks of proteins. Thus, a supramolecular host molecule that has a high affinity for specific residues or sequences would be beneficial for sensing or manipulating biological systems. In this section, the biological molecules that contain recognition motifs for CB[*n*] are discussed with an emphasis on amino acids, peptides and proteins. Then, the biological applications of such CB[*n*] biomolecule recognition, including control of enzyme activity and non-covalent protein modification, are described.

7.1.1. *Recognition of amino acids, peptides and proteins by CB[6]*

CB[6] forms only modest exclusion complexes with various amino acids, with very little differentiation between them (all $K_a < 10^3$ M^{-1}),[15,416] with the exception of lysine, the side chain of which can be threaded into CB[6].[41] The presence of a carboxylate, however, makes it a weak binder (8.7×10^2 M^{-1}).[72] As such, CB[6] has limited application to recognize amino acids. To address this limitation, a fluorescent analogue of CB[6] that has aromatic walls and is also large enough to encapsulate aromatic amino acids has been developed.[417] This CB[6] analogue favours aromatic residues because it can make π–π interactions with the aromatic side chains.

Lysine-containing peptides and proteins are recognised by CB[6] (e.g. Lys–Ala, $K_a = 1.6 \times 10^4$ M^{-1}).[338] Such recognition has allowed CB[6] to bind to lysine residues in proteins, and the binding affects the proteins' solubility.[418] The phase-transfer effect has been exploited to control the growth of amyloid fibrils at the nanoscale. This work represented the first example of kinetic control of fibril assemblies by use of a supramolecular host molecule. CB[6] binds weakly ($K_a \sim 10^2$–10^3 M^{-1}) to most other, non-lysine containing peptides, with little discrimination.[172]

7.1.2. *Recognition of amino acids, peptides and proteins by CB[7]*

CB[7] can encapsulate the side chains of various hydrophobic and aromatic amino acids, namely phenylalanine, tyrosine and tryptophan. Among them, phenylalanine was found to be the best binder with an affinity of around 10^6 M^{-1}.[419] The main reason for this is that it has good size complementarity with the cavity and there is strong hydrophobic interactions between the interior of the cavity and the amino acid side chain.[186] CB[7]'s affinity for phenylalanine is retained when phenylalanine is part of a peptide sequence. For example, CB[7] binds well to short, phenylalanine-containing dipeptides. Peptides with N-terminal phenylalanine had a higher affinity ($K_a \sim 10^6$ M^{-1}) compared to those where the phenylalanine was not located at the N-terminus.[61]

A different study revealed that CB[7] could discriminate between diastereomeric peptides (up to 8-fold for $^+$H$_3$N–*L*–Phe–*L*–Leu versus $^+$H$_3$N–*L*–Phe–*D*–Leu) when a chiral molecule is bound first.[61] The principal finding of this study was that chiral recognition is possible inside the achiral cavity provided that a chiral guest is already bound, meaning that diastereomeric ternary complexes can be formed. In the case of a dipeptide, the chiral centres are connected by a covalent bond; the first chiral residue binds (e.g. Phe), and this dictates the preferred stereochemistry at the second position.

CB[7] has been shown to bind to human insulin through the N-terminal phenylalanine on its *B*-chain.[24] ITC revealed that CB[7] makes a complex with human insulin with an affinity of 10^6 M^{-1}; substitution of this position by mutagenesis abolished binding. An X-ray crystal structure showed that not only does CB[7] bind at the phenylalanine but it also causes the preceding residues to unravel to accommodate CB[7]. The

N-terminal-specific binding is very useful because there is only one N-terminus per protein and, oftentimes, it can unfold more easily than elsewhere on a protein. The specific binding to the N-terminus has been exploited for the enhancement of peptide signals in mass spectrometric (MS) analysis.[187] When CB[7] is added to a digested protein, it can bind N-terminal phenylalanine. As a consequence, proton transfer is facilitated between basic residues under MS conditions, thereby giving rise to alternate fragmentation pathways during collision-induced dissociation (CID). The addition of CB[7] allowed more residues to be detected, which is termed increased "sequence coverage" in the proteomics field. Similarly, the addition of CB[7] to N-terminal peptides can be used to inhibit the action of a protease. Exoproteases sequentially cleave the terminal residue of a peptide from the N-terminus to the C-terminus. When the N-terminal phenylalanine is revealed, CB[7] binds strongly to that residue and arrests the cleavage of the protein.[188]

There are also potential therapeutic applications of CB[7]–phenylalanine binding behaviour. Phenylalanine is a critical residue in amyloid fibril formation — a key process in the progression of Alzheimer's disease. Clustering of phenylalanine residues mediated by hydrophobic interactions is thought to be one process that directs fibril formation. Therefore, sequestration of the phenylalanine residues by CB[7] can be used to inhibit fibril formation.[420]

While the binding to phenylalanine is quite strong, it is not as strong as some other CB[7]–guest pairs, so efforts have been made to improve this interaction. Various phenylalanine analogues that were modified at the *p*-position of the phenyl ring have been reported.[421] These studies found that *p*-methylaminophenyl alanine at the N-terminus of peptide binds to CB[7] 20–30-fold higher than any canonical amino residue in a peptide ($K_a \sim 10^{10}$ M^{-1}). Such non-canonical amino acid residues may be useful as minimal affinity tags in future applications, such as imaging and enrichment of proteins.

7.1.3. *Recognition of amino acids, peptides and proteins by CB[8]*

CB[8] itself is too large to recognise amino acids on its own, but CB[8] binary complexes with a π-acceptor guest such as methyl viologen (MV) recognise amino acids that act as π-donors such as tryptophan.[21]

Urbach *et al.*[60] further investigated and quantified this behaviour and found that CB[8]–MV bound tryptophan (K_a ~ 10^4 M^{-1}) with 8- and 20-fold enhanced selectivity over phenylalanine and tyrosine, respectively. No other amino acid seems to bind, and the selectivity was dictated by the hydrophobicity of the amino acid and its ability to make van der Waals interactions with the cavity.[422] Another crucial factor is the positive charge on the amino acid, whereas steric bulk is not that important; these insights were determined by ITC experiments with tryptophan analogues.[60] Kaifer *et al.*[423] demonstrated that tryptophan recognition could also be confirmed by fluorescence quenching of a CB[8]-2,7-dimethyldiazaphenanthernium (DAP). DAP is inherently fluorescent, but its fluorescence is quenched when tryptophan binds, and forms a ternary complex since it completes a charge transfer complex.

Since cationic tryptophan derivatives can bind CB[8]–MV, peptides with N-terminal tryptophan should also be able to bind this complex with some selectivity. Binding studies with various tryptophan-containing tripeptides revealed that this is indeed the case. The peptide H_2N–WGG (K_a ~ 10^5 M^{-1}) bind the strongest followed by H_2N–GWG (6-fold less) and H_2N–GGW (40-fold less).[60] In each case, the binding was enthalpically favourable and entropically unfavourable. The high affinity of the complex compared to that of other small molecule–biomolecule recognition is noteworthy, and the optical properties of the system i.e. quenching of the tryptophan's fluorescence, is useful as an intrinsic detection mechanism. This recognition behaviour can be exploited to prepare surfaces that can capture tryptophan-containing peptides and then release them in response to an electrochemical stimulus.[424] Another guest of CB[8] that can be used for peptide recognition is tetramethylbenzobis(imidazolium) (MBBI), which, unlike MV, has intrinsic fluorescence, which allows detection of second guest binding; the previous studies relied on the fluorescence of tryptophan. MBBI forms similarly strong complexes with CB[8] and N-terminal tryptophan peptides, analogous to the MV complexes.[425] The CB[8]–MBBI has been used to induce and report on the binding of specific peptide sequences.[426] Based on the previous studies, it was known that an N-terminal aromatic residue was crucial for strong binding. The intrinsic fluorescence of MBBI, however, meant that the presence of tryptophan was not necessary to determine binding, so a wider range of peptide sequences could be surveyed. A library of over 100 peptides was surveyed and it was found that the N-terminal residue is not the only important one for binding; the second and third amino acid also modulate the binding significantly. This contrast is shown starkly in the peptides Tyr–Leu–Ala and Tyr–Ala–Leu, the former is a nanomolar binder ($K_d \sim$ 7.2 nM), whereas the latter is a micromolar binder ($K_d \sim 34 \ \mu$M) with four orders of magnitude difference. Careful ^{1}H NMR studies revealed that there were interactions between CB[8] and the first two amino acids (tyrosine and leucine) in the high-affinity peptide, but only with tyrosine in the lower-affinity peptide. Peptides with a terminal phenylalanine or tryptophan are also strong binders, but they displace MBBI to form a 1:2 complex with CB[8].

The large volume of CB[8]'s cavity can accommodate two tripeptides in a 1:2 (host–guest) fashion. Among the tripeptides investigated, it was found that only Phe–Gly–Gly (H_2N–FGG, $K_a \sim 10^{11}$ M^{-2}) and Trp–Gly–Gly (H_2N–WGG, $K_a \sim 10^9$ M^{-2}) dimerised in the CB[8] cavity; the dominance of the 1:2 species in sub-stoichiometric mixtures indicated positive cooperativity (Figure 7.1).[63] The high binding affinity for FGG was due to an additional enthalpic contribution, which seemed to result from the staggered face-to-face stacking of the phenylalanine side chains. CB[8]-mediated peptide dimerisation is not limited to N-terminal residues: pentapeptides with phenylalanine as the middle residue could form

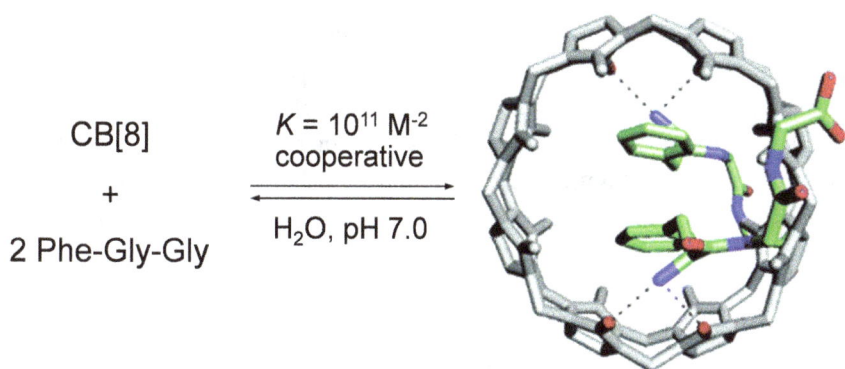

$$CB[8]$$
$$+$$
$$2 \text{ Phe-Gly-Gly}$$

$K = 10^{11}$ M^{-2}
cooperative

H_2O, pH 7.0

Figure 7.1.   The cooperative dimerization of two Phe–Gly–Gly peptides inside CB[8]. (From Ref. 63)

homodimers by encapsulating the phenylalanine side chains inside CB[8].[427] The peptides surveyed differed in the amount of steric hindrance and the polarity of the charge, which suggests that such homodimerisation may be generally applicable.

The key point to take away from this section is that CB[8] has exquisite sequence selectivity for certain peptide sequences. These sequences have been discovered mainly by the efforts of Urbach's group. The affinities and sequence specificity of these complexes are generally higher than those for other synthetic receptors. These high-affinity CB[8]-binding epitopes present an opportunity to use CB[8] to manipulate and label natural proteins. A crucial point to consider here is that these epitopes are made from natural amino acids, which can be easily incorporated into recombinant proteins. The straightforward introduction of these epitopes may expedite their transition to real biological applications.

7.1.4. *Non-covalent protein modification*

In chemical biology, there has been considerable interest in covalently modifying peptides and proteins with the aim of controlling the proteins' functions or imaging them. Nature, on the other hand, can not only control proteins covalently (e.g. through post-translational modifications) but also non-covalently (e.g. through protein–protein interactions, protein–ligand interactions, among others). Chemists have struggled to replicate nature's non-covalent and reversible control over biological systems. CB[8]'s ability to form ternary complexes including specific peptide sequences presents an opportunity to modify proteins non-covalently and reversibly.

Scherman *et al.* reported the first example of CB-mediated non-covalent protein modification.[428] They sought to non-covalently label a protein with PEG (Figure 7.2). PEGylation of therapeutic proteins shows promise in therapeutics since the PEG groups increase the proteins resistance to metabolic enzymes, increase their solubility and extend their clearance time *in vivo*. They first covalently modified BSA using a maleimide linked to a naphthyl group, they did this site selectivity by exploiting the fact that BSA has only one cysteine residue in the reduced form. They then introduced MV–PEG and CB[8] leading to the formation of a protein–PEG conjugate, which was confirmed by various analytical methods including DOSY–NMR and UV-vis spectroscopy.

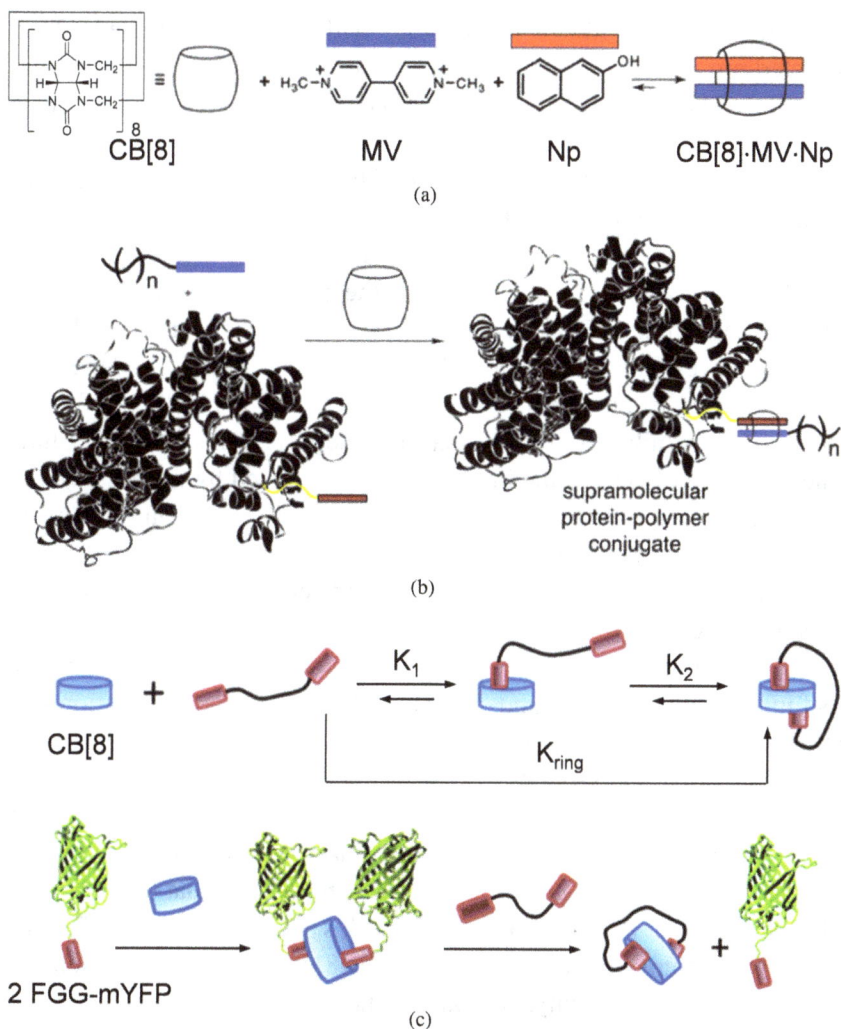

Figure 7.2. (a) CB[8] couples together methyl viologen and a naphtyl derivative. (b) Naphtyl-functionalised BSA-can be non-covalently linked to methyl viologen terminated polymer. (From Ref. 428) (c) Proteins with an N-terminal FGG sequence can dimerise in the presence of CB[8]; the process is reversible by addition of a competitor. (From Ref. 431).

Brunsveld *et al.*[86] used a similar strategy to form a protein complex between two different proteins, artificially. They modified each protein at the C-terminus using the native chemical ligation strategy: one protein was modified with MV and the other with a naphthyl moiety. They selected proteins that were FRET pairs so that the dimerisation was auto-reporting. The non-covalent dimerisation of the proteins was confirmed by the occurrence of FRET between the two proteins when the proteins were treated with CB[8]. To exploit CB[8]'s ability to facilitate the dimerisation of N-terminal FGG sequences, a genetically engineered protein pair that included an FGG motif at their N-terminus was pre-pared. The addition of CB[8] allowed the proteins to dimerise.[429] The homodimerisation of FGG-tagged yellow fluorescent protein (YFP) has also been investigated; it was found that first dimers formed, and then two of these dimers come together to form a tetramer.[430] This dimer of dimer formation is similar to the way that tetrameric proteins form in nature. The assembly of such homodimers could be reversed by the use of a lin-ear small molecule that has an FGGG motif at both ends (Figure 7.2(c)).[431] The small molecule displaces the proteins and forms a ring. This mole-cule was more efficient at dissembling protein dimers than adding a simple peptide.

In nature, the main purpose of protein complexation is regulation of its function. With this in mind, CB[8]-mediated protein dimerisation sys-tems have been used to switch on an enzyme. In these cases, the addition of CB[8] caused the proteins to dimerise and thereby turn on their func-tion. The first example used the protease, Caspase 9.[97] In dilute solutions, Caspase 9 is normally in the monomeric, inactive state. Its active form is dimeric, and the crystal structure of the dimer revealed that the N-termini of each monomer are close to each other. This suggested that modification of the N-terminus to FGG would allow CB[8] to induce dimerisation and turn on proteolytic activity. Not only could the protein activity be turned on by CB[8], the system was completely reversible by addition of a com-petitor peptide. Such reversible control of the activity of a mature protein is difficult using conventional methods. Here, the protein is controlled by addition of a three-residue sequence which does not hamper biological activity of the proteins. The second example is a split luciferase system.[432] Split luciferase systems are a tool to report protein–protein interactions. Intact luciferin is luminescent; the luciferin fragments weakly bind to each other but not strongly enough to switch on luminescence.

CB[8]-mediated dimerisation can be used to stabilise the protein complex, which causes the luciferin to switch on. The principle of using a small supramolecular host to stabilise weak or low concentration protein complexes may be useful in monitoring transient events in signalling pathways. Furthermore, they may find applications in biohybrid systems that artificially regulate protein functions or signal transduction pathways.

As discussed above, there are specific peptide sequences that CB[8] can recognise. The CB[8]–FGG interaction has also been used to make protein-based assemblies, such as nanowires synthesised by Liu *et al.*[433] By taking advantage of a naturally dimeric protein (glutathione S-transferase, GST) and functionalising the N-terminal ends with FGG peptides, polymeric nanowires could be constructed by the addition of CB[8] (Figure 7.3). Such a system has been used as the basis of functional protein nanowires; other proteins can be inserted in between GST and the FGG motif to endow function. For example, antioxidant proteins have

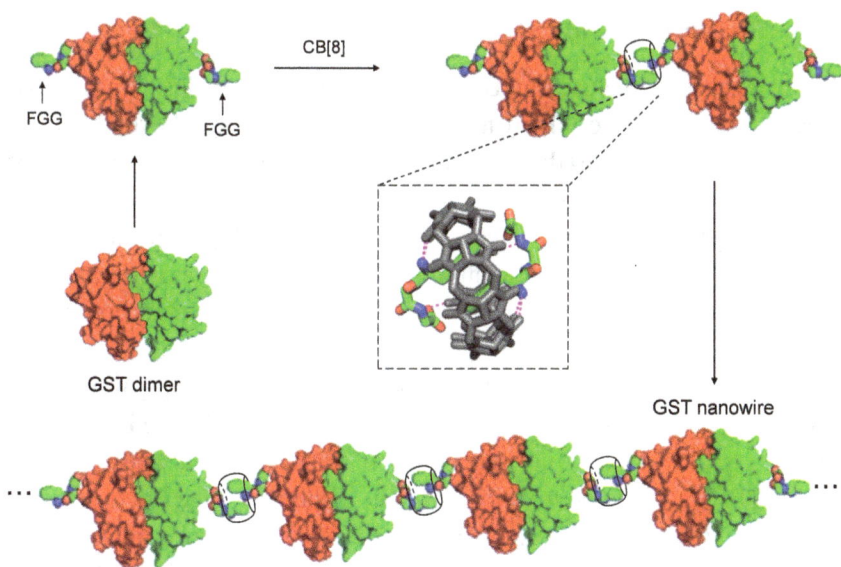

Figure 7.3. A CB[8]-linked protein nanowire. (Modified from Ref. 433)

been fused to the GST to give the assemblies antioxidant activity,[433] and a Ca^{2+}-dependent protein has been added to allow the assembly to behave like a nanospring.[434] The host–guest interaction can also be used to organise the assembly prior to covalent crosslinking of the assembly.[435]

7.2. CB-based Drug Delivery and Imaging

Various supramolecular host molecules have been used as drug carriers that improve solubility, stability and bioavailability of drug molecules by encapsulation. Among them, CB[*n*] have become key players in this area because of their thermal, chemical and *in vivo* stability. This section serves to discuss concepts involved in drug delivery using CBs, including toxicity profiles, release mechanisms, and pertinent examples of their uses. Furthermore, CB-based materials that have been used as targeted drug delivery vehicles are also discussed. Nau has recently written a book chapter[436] detailing the various drugs that have been encapsulated in CB[*n*]; readers are directed there for a broader look at the field.

7.2.1. *Toxicity profile of CB[n] compounds*

Before discussing the use of CB[*n*] as drug delivery vehicles, a reasonable question to ask is: Do CB[*n*] have an inherent toxicity? Several groups have investigated both the general and the organ-specific toxicity of CB[*n*]. The cytotoxicology of CB[5], CB[7] and some acyclic variants towards liver and macrophage cells was evaluated using a typical cytotoxicity assay using 3-(4,5-dimethylthiazol-2-yl)-2,5-diphenyltetrazolium bromide (MTT), which revealed that > 90% cells survived at concentrations up to 1 mM.[82]

In another study, the toxicity of both of CB[7] and CB[8] was evaluated *in vitro* and *in vivo*.[81] The *in vitro* assays with CB[7] were in line with those above, demonstrating that CB[7] is not significantly toxic below 1 mM. CB[8] showed no toxicity within its solubility range (up to 20 μM). The *in vivo* toxicity was evaluated by intravenous and oral administration in a mouse model. Acute toxicity, defined here as greater than 10% body weight loss, was not observed when CB[7] was administered intravenously at 250 mg kg^{-1}; this value is far higher than needed for drug delivery. CB[8] was not tested intravenously due to its poor solubility. CB[7] and CB[8] were administered orally at 600 mg kg^{-1} as a single dose and appeared to be safe.

To investigate organ-specific toxicity, Wheate *et al.*[437] performed *ex vivo* neuro-, myo-, and cardiotoxicity studies with CB[6], CB[7] and an acyclic analogue and compared the results to that of β-cyclodextrin. A mouse sciatic nerve was exposed to 1 mM solutions of the four macrocycles; conductivity measurements showed no significant neurotoxicity. Biventer cervicis were extracted from chicks, and their responsiveness to chemical or electrical stimuli was measured before and after exposure to 0.3 mM of each of the four containers. Only a small change in muscle contraction strength was observed with CB[6]; hence, it can be assumed to have low mytotoxicity. The acyclic version and cyclodextrins were not mytotoxic. However, CB[7] seems to have a greater effect on muscle contraction. The authors suggest that CB[7] may bind to nicotinic receptors, resulting in the mytotoxic effects. Heart atria were extracted from rats and treated with each container at 0.3 mM and the rate of contraction (beats per minute) of the heart was measured. Each CB showed significant cardiotoxicity, with the acyclic CB being the most toxic. Although myto- and cardiotoxicity were observed, this may be because the concentrations and the direct exposure were not representative of the drug delivery situation. A study on the toxicity of CB[7] has been conducted with live zebrafish.[438] In this study, cardiotoxicity was observed, as the above study did, albeit at concentrations above 500 μM. Liver toxicity was also examined which revealed that the liver seems to be able to tolerate CB[7] below 750 μM. Taken together, CB[7] toxic effects are not apparent at typical therapeutic concentrations.

Acyclic CBs have also been evaluated for toxicity.[150] Examination of their cytotoxicity with the key cell types affected by toxic compounds, namely, liver, kidney and immune cells, revealed that they were safe up to 10 mM. They determined the maximum tolerated dose in mice to be 1,230 mg kg^{-1}; furthermore, these mice recovered their weight to the levels of the control group over time. While all these results are promising, more studies are needed before applying it into real animal-level applications. In addition, it remains to be seen if the CBs and related compounds are safe for larger animals and, ultimately, humans.

7.2.2. *CB[n] and related compounds as drug containers*

Encapsulation of a drug inside a CB[n] container affects the physiochemical properties of the drug molecule. For example, amine-containing drugs undergo a pK_a shift (up to 5 units); this has been exploited to activate prodrugs.[165] The drug can also be protected from biomolecules and its solubility improved. While several different classes of drugs have been shown to be guests of CB[n], only anticancer drugs are described here because they best embody the advantages of CB[n] drug containers and there are data ranging from test tube to *in vivo* level. More details can be found in Nau's book chapter which gives a more complete account of CB[n] in drug delivery.[436]

Platinum-based complexes are a widely used class of anticancer agents which inhibit replication of DNA by binding in-between DNA base pairs. Some Pt complexes have hydrophobic ligands attached to the metal centre through which the drug can be, at least partially, encapsulated in CB[n]. In 2004, Wheate *et al.*[439] first demonstrated the encapsulation of a dinuclear Pt anticancer compound in CB[7]. They showed that CB[7] protects the Pt centre from biological thiols, which can displace ligands around the Pt centre, making the drug inactive. Furthermore, by use of different sized CB[n], the release rates and cytotoxicity can be tuned.[440] Among anti-cancer compounds, oxaliplatin binds to CB[7] particularly strongly (10^5 M^{-1}).[441] An X-ray crystal structure of the complex revealed that the cyclohexyl group was encapsulated and the amine groups were in the same plane as the portal. Moreover, the encapsulated drug was more stable than the free form in the presence of methionine and guanosine.

Cisplatin@CB[7] was evaluated for its ability to reduce the growth of an ovarian tumour in mice.[93] The drug complex reduced the tumour doubling time 1.6-fold. Interestingly, no such effect was observed in an *in vitro* model. The reason for this is that cisplatin is released slowly from CB[7], resulting in a sustained therapeutic effect, whereas free cisplatin is consumed more quickly by degradation processes; they elucidated this by measuring cisplatin content in the bloodstream over time.

A method to target the delivery of anti-cancer complexes has been described.[98] A biotin-conjugated CB[7] (BT–CB[7]) was used to target cells that overexpress biotin receptors. BT–CB[7] containing oxaliplatin showed 200-fold greater cytotoxicity to cancer cells than free oxaliplatin does, as judged by a cell viability assay. The reason for the improved efficacy appears to be the targeting effect, which if it could be replicated *in vivo* would allow for a reduction in the amount of oxaliplatin required for cancer treatment. Because of the significant amount of endogenous biotin and biotin receptors, it may not be a specific targeting method. Other targeting groups such as peptides may solve the issue.

Temozolomide (TMZ) is a powerful chemotherapeutic agent for the treatment of glioblastoma multiforme (GBM). The molecule is a prodrug which converts to an active form when it is exposed to physiological conditions. The active form, however, is unable to penetrate the blood–brain barrier (BBB) but the pro-drug is. Thus, reducing the conversion of the prodrug to active form before it penetrates BBB is important to enhance TMZ efficacy to GBM. To slow the rate of conversion, TMZ was encapsulated in CB[7], which protects TMZ from degradation (Figure 7.4).[442] Cell viability assays with GBM cells revealed that the compound

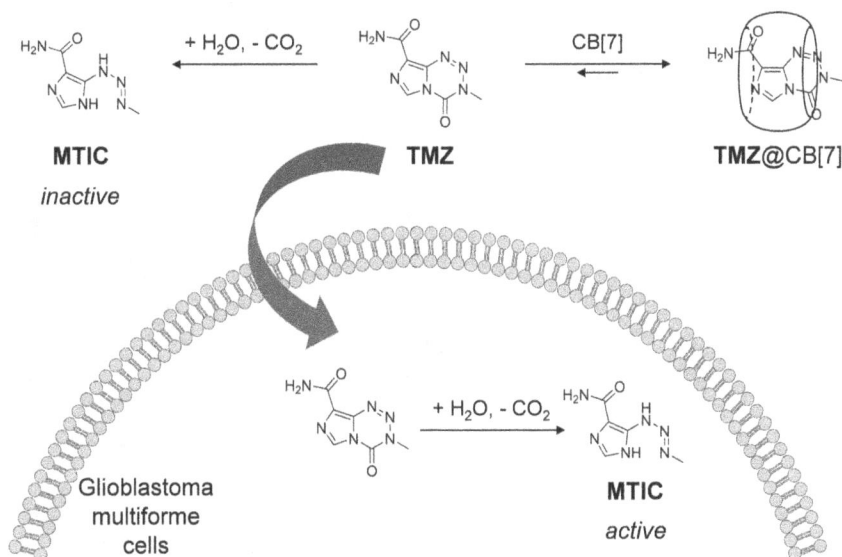

Figure 7.4. CB[7] protects TMZ from premature decomposition. Decomposition inside the cell affords the active compound. The TMZ@CB[7] complex is labile and does not enter the cell. (From Ref. 442)

administered with CB[7] was much more active than the free drug, indicating that CB[7] protects TMZ from degradation in physiological conditions.

The solubility of new drug candidates is a problem in drug discovery; in fact, 40–70% are not soluble enough to be formulated alone. CB[n] can help solubilise a range of drug molecules; however, this not a general approach because of the size constraints of the container. To address this, Isaacs *et al.*[150] have synthesised flexible acyclic CB-based containers functionalised with sulphonate groups as general solubilising agents (Figure 7.5). When the drugs were complexed with the containers, 36–2750-fold enhanced solubility was observed compared to that of the free form. The activity of paclitaxel was compared to that of paclitaxel encapsulated in an acyclic CB; the increased concentrations may provide a greater number of drugs to cancer cells.

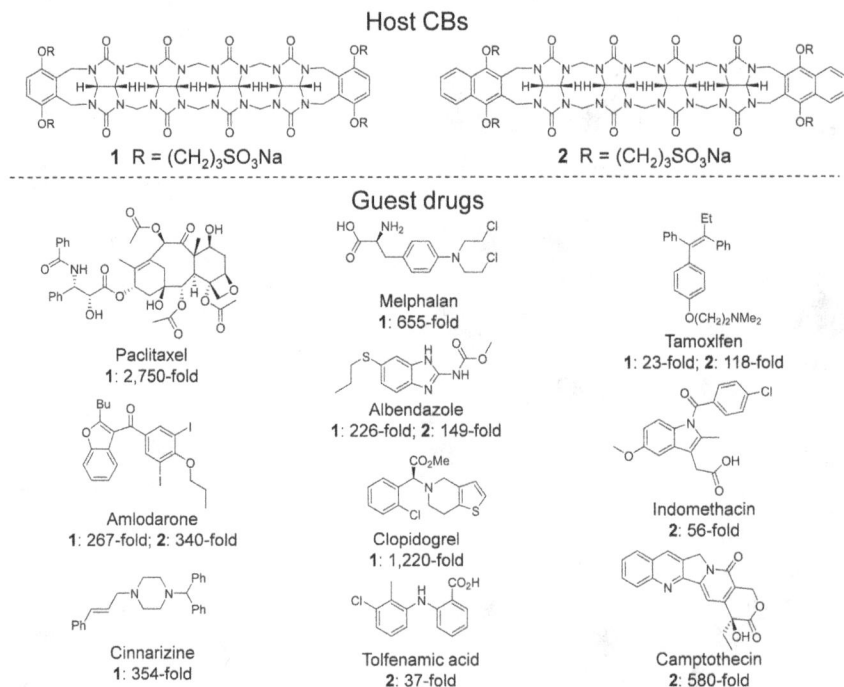

Figure 7.5. Acyclic CB[n] containers and their solubility enhancement of selected drug molecules. (From Ref. 150)

There have been some investigations into the use of CB[n] in drug formulations. Microcrystalline CB[6] has been formulated into a 50% w/w with the other components that make up a typical oral tablet.[443] The tablets had appropriate properties including mechanical strength and disintegration in simulated gastric or intestinal fluid. CB[6] has been prepared as eye drop formulation and a topical cream[444]; however, they do not permeate the cornea or the skin efficiently, which limits their use to localised topical treatments. Polymorphism is an important consideration in solid state formulations of drug molecules. Different solid state structures (polymorphs) are considered separate chemical entities by the FDA and other regulatory bodies because they can have different physiochemical properties. Some drug molecules can undergo a transition from one polymorph to another while sitting on the shelf. CB[7] has been shown to suppress polymorphism for several drug molecules, suggesting their use as a stabiliser in formulations.[445] On the other hand, CB[n] themselves also have highly variable solid state structures depending on the synthesis method and (anti)-solvents of crystallisations. In particular, the water content is highly variable ranging from 0.5% to 18% w/w.[446] Another issue is CB[n]'s compatibility with other reagents in the formulations since the drug and the macrocycle are unlikely to be the only components in the formulation. While this area needs more thorough investigation, what is known in this nascent area, and of CB drug formulation more generally, has been reviewed recently by Wheate.[446]

7.2.3. Drug release mechanisms

Many host–guest complexes associate and dissociate on the rate of seconds or faster, ensuring a dynamic equilibrium appropriate for drug release. However, efforts have been devoted to shifting the equilibrium further towards dissociation with triggered release on-demand. Since drug molecules are bound to CB[n] non-covalently, they can be released spontaneously by dissolution from the complex. For example, the drug molecule, albendazole, was shown to exist in the free form within seconds.[447] A similar study showed that dinuclear Pt^{2+} and Ru^{2+} complexes dissociated from CB[10] within hours.[200] It is worth noting that binding affinities alone are not a useful guide to how available an encapsulated drug will be. This is because of the tight, constrictive binding that guests inside CB[n] experience. However, slow and sustained release of a drug over time can improve its efficacy (see the cisplatin@CB[7] example above). Isaac *et al.* reported that the rates of dissociation for their acyclic CB analogues are faster than those of cyclic CBs, due to their less constrictive binding.[150]

Several research groups have exploited the reversible binding of CB–guest complexes to release the drugs in response to specific conditions. For example, salt effects have been exploited to weaken the binding of CB[7] to guest and encourage the guest to move into a hydrophobic pocket of a protein.[448] In a more targeted process, Pischel *et al.*[449] have demonstrated that guests can be released with a photo-induced pH jump. The authors encapsulated Hoechst 3325, an antiparasitic drug with a pK_a 5.5, inside CB[7], in the presence of a photo base (Malachite green). Upon UV irradiation (300 nm), hydroxide is released from Malachite green to initiate a localised pH jump, which causes the drug to be released from CB[7]'s cavity. A similar principle to acidify the media upon UV irradiation, using *o*-nitrobenzaldehyde as the photoacid showed displacement of amines (by extrapolation amine-containing drugs) from CB[7] by a protonated fluorescent dye which has higher binding affinity to CB[7] when protonated.[450] The pH dip from 8 to 5 is enough to cause the guest exchange; interestingly, changing the pH of the solution without the dye did not cause the amine to be released. These approaches suggest that drugs with appropriate pK_a can be released by a photo-initiated pH jump on demand. Although, this was demonstrated with UV light which is known to be harmful to cells, near IR or two photon irradiation methods may help to extend this strategy into real biomedical applications.

7.2.4. *CB-based materials as theranostic materials*

Vesicles have experienced a great upsurge of interest in recent years because of their potential in drug delivery and for the creation of biomimetic systems because of their resemblance to a cell membrane. Vesicles are formed from amphiphilic molecules which self-assemble into cell-like nanostructures. Kim *et al.*[452] synthesised a derivative of CB[6] that could

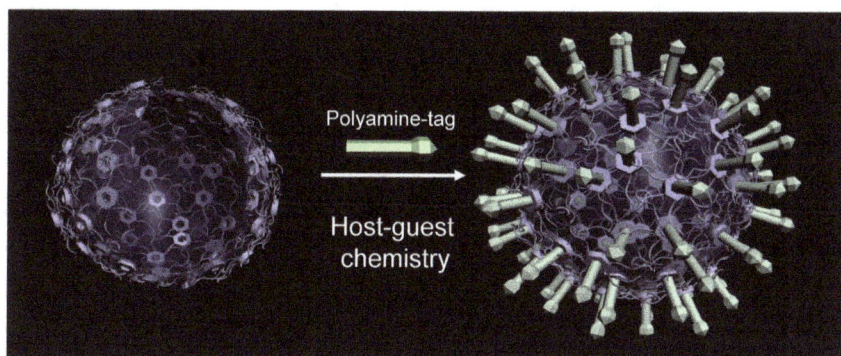

Figure 7.6. Illustration of the surface modification of vesicle through host–guest interactions between CB[6] and polyamines. (From Ref. 451)

form a vesicle; the surface of the formed vesicle was non-covalently tailorable (Figure 7.6). In their initial study, they demonstrated that the surface could be modified with fluorescent dyes and sugar moieties using strong host–guest interactions between CB[6] and polyamines such as spermine and spermidine (K_a ~ 10^{10}–10^{12} M^{-1}).[175] It has also been suggested that these vesicles could be functionalised in such a way that it could be used for targeted theranostics. In a follow-up study, these vesicles were used as the scaffold of a multifunctional drug delivery platform.[453] They modified the decoration of the CB[6] to make it reduction sensitive, and hence the vesicle could be disassembled by reducing environments such as the cytosol, as a drug release mechanism. The anticancer drug, doxorubicin (Dox), was loaded into the vesicles and the vesicle surface further functionalised with imaging probes and a targeting ligand to activate receptor mediated endocytosis. After internalisation, the vesicles were visualised under a fluorescence microscope, and were found to be more cytotoxic than free Dox. Another CB[6]-derivative can self-assemble into a nanoparticle which has a non-covalently tailorable surface. This nanoparticle was used as a drug delivery vehicle by non-covalently modifying the surface with spermidine-tagged drug molecules.[454] This strategy was especially helpful for delivery of hydrophobic drugs. Using receptors that trigger endocytosis allowed the drug loaded nanoparticles to be internalised into the cell.

Self-assembling, polymer-based theranostic systems have been developed by preparing a hyaluronic acid-functionalised on the side chain with CB[6].[455] Various functional tags conjugated to spermidine were then introduced to make a multifunctional polymer including bioimaging agents, targeting aptamers, bio- and chemical pharmaceuticals. A model system comprised of a targeting peptide-conjugated spermidine and a fluorescent-probe-conjugated spermidine was used to show targeted delivery of the material into cells, and the stability of the complex was demonstrated *in vivo* in a mouse model. The key strength of this method is the facile self-assembly allowing simple on demand tuning of the desired properties.

The polymer nanocapsules that were introduced in Section 6.4 have several properties that can be exploited for a drug delivery and imaging applications. For example, they can swell and compress in response to solvent.[456] The capsules are synthesised in methanol, but when the solvent is exchanged to 10% methanol–water, the nanocapsules swell and consequently the permeability of capsules increases; switching the solvent back to methanol compresses the capsule. This behaviour may be used to encapsulate drugs. Another approach to encapsulate a drug is to synthesise the nanocapsules in the presence of the drug. The drug can be released passively (slow leaking out of the cargo) or actively by changing the chemistry of the linker between the CB[6] units.[457] For example, incorporating a disulphide allows the triggered release of the drug molecules in response to a reductive environment. The surface of the nanocapsules can also be non-covalently functionalised with various functional tags such as targeting ligands and imaging probes by exploiting the host–guest interactions between CB[6] and spermidine. A recent example used CB[6] nanocapsules as a platform for cancer-targeted multimodal bioimaging in animals (Figure 7.7).[396] The nanocapsules were simultaneously functionalised with a near-infrared dye, a PET-tracer and a cancer-targeting peptide. The key to the success of this strategy is the robust nature of the non-covalent modifications. The host–guest interaction remained strong in the complex and hydrodynamic environment of an animal's blood stream. Such a non-covalent approach would not be possible with other supramolecular hosts because of the weaker binding affinities, especially in physiological conditions.

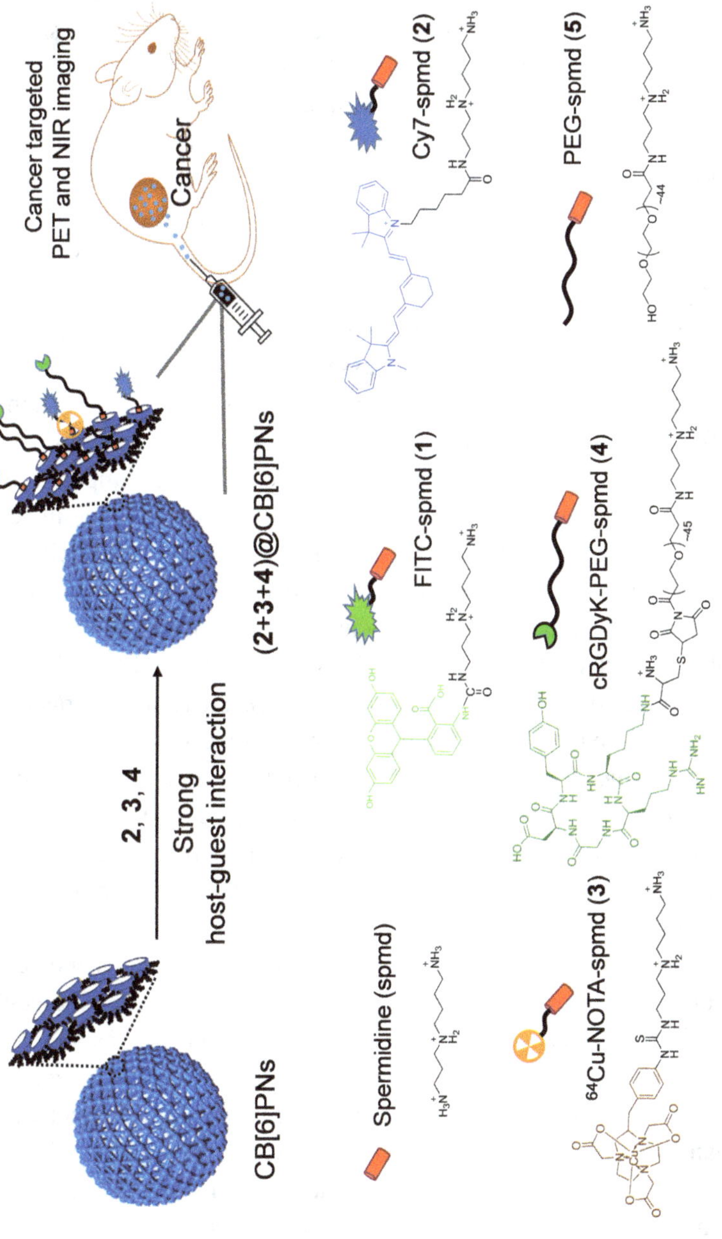

Figure 7.7. Facile surface modification of CB[6]–PNs using the strong host–guest interactions of CB[6] with spmd-functional tags in a non-covalent and modular manner and the use of CB[6]–PNs as a robust and multifunctional platform for multimodal imaging of a cancer-bearing mouse. (From Ref. 396)

CB[8]-based vesicles have also been employed as stimuli-responsive drug delivery vehicles. Scherman *et al.*[458] describe the preparation of supramolecular peptide amphiphiles double layer vesicles. They showed that these vesicles could be taken into a cell and the addition of competing guests caused the vesicle to disassemble since the CB[8]-ternary complex was broken. Depending on the competing molecule used, a cytotoxic effect was observed. When an Np-derivative was added, only the pyrene-peptide conjugate was released and a single-layer vesicle formed. However, when an adamantylamine was added, the whole vesicle dissembled resulting in cytotoxicity. The disassembly of the vesicle was accompanied by an increase in fluorescence as the fluorophore is released from CB[8].

Micelles are another class of self-assembled nanostructures, and like vesicles, can be formed from CB-based amphiphiles. Such micelles can be formed from stimuli-responsive block co-polymers, for example from temperature- and pH-sensitive polymers.[459] Low pH, high temperature or the presence of a competing guest caused the disassembly of the micelle. These micelles were loaded with Dox and tested for their cytotoxicity. The efficacy was compared to that of free Dox. Individually, the release mechanisms did not reach the same efficacy as free Dox; however, the combination of competitive guest and low pH showed comparable efficacy to free Dox. A related example described a sugar-sensitive insulin releasing system. In this case the pH-sensitive group was replaced with one that was susceptible to glucose.[460] While this was an innovative approach to making a biomimetic sugar-responsive insulin delivery vehicle, typical physiological quantities of glucose were insufficient to release the protein. A similar strategy where micelles are disassembled either by oxidation or treatment with competitive guests has also been reported.[461]

Mesoporous silica has been used as a carrier for various molecular cargo, including drug molecules. To control the release of cargo molecules from mesoporous silica, pseudorotaxanes have been utilised as a molecular valve by using stimuli sensitive gating mechanism. Zink *et al.* have been the pioneers in this area.[462] The basic principles and the activation mechanisms of these nanovalves were discussed in Chapter 4. Although there appears to be potential for these materials in drug delivery, there have been few examples of such a system being used for drug delivery. For example, Zink *et al.*[463] reported a mesoporous silica that was

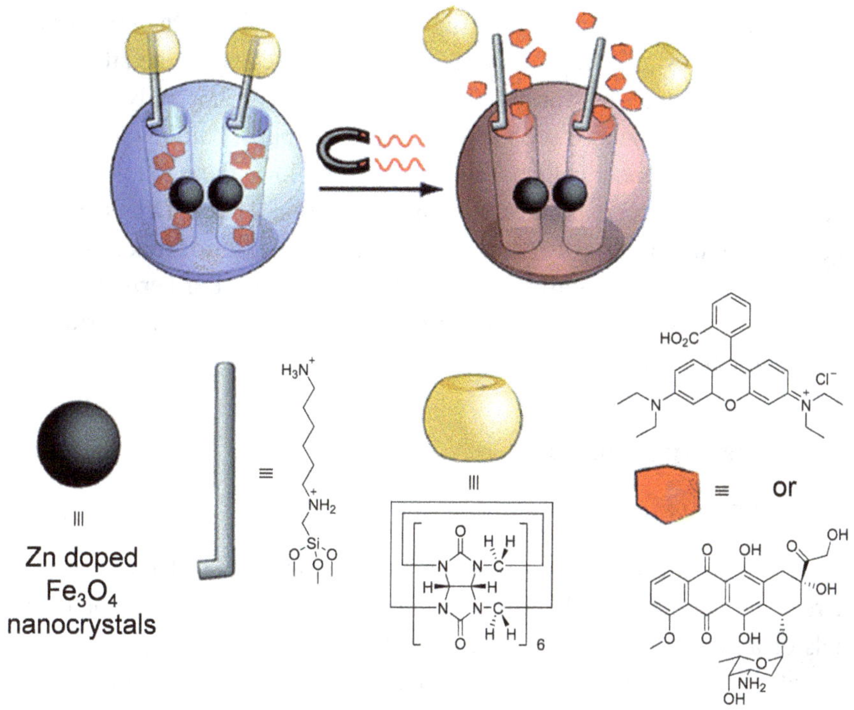

Figure 7.8. Nanogated mesoporous silica nanoparticles. Exposure to a magnetic field generates local heat which causes the CB[6] to unthread from the "stalk", thereby releasing the cargo, in this case Dox. (From Ref. 462)

impregnated with Zn-doped Fe_3O_4 nanocrystals to sensitise them to magnetic fields, and its pores were gated with a CB[6]-pseudorotaxane to entrap and release Dox (Figure 7.8). The drug was released by applying an oscillating magnetic field, which caused the temperature around the drug delivery vehicle to rise. The increased local temperature caused dethreading of CB[6] from the stem to open the valve, which allowed the drug molecules to be released.

To summarise the broad field of CB-based drug delivery, many drug molecules have been shown to be guests of CB[n] (n = 6–8). CB[7]'s aqueous solubility, high affinity for various drugs, and safe toxicology profile has led to it being used as a drug container. Simple dilution of CB–drug complexes, in a blood stream, for instance, is usually enough to release the drug molecule from the container. CB-based nanomaterials (such as vesicles, nanocapsules and micelles) can act as targeted theranostic multifunctional platforms. Such platforms can not only be non-covalently functionalised with active targeting groups, but they inherently passively target cancer cells because their size allows them to take advantage of the enhanced permeability and retention effect. Drugs can be loaded into the interior of these platforms and the exterior functionalised with imaging agents to allow theranostic treatments. Furthermore, drug release can be stimuli-responsive to physiological environments, such as reductive conditions.

7.3. Other Applications

7.3.1. *Ion channels*

Ion channels are crucial components of cells that control ion transport across membranes. Synthetic ion channels that replicate natural ion channels or add new functionality are highly sought. Considering the properties of CB[6], the dimensions of the CB[6] portal fits quite well with the size of the K^+ but is too large for Na^+, therefore CB[6] was examined as an artificial ion transporter with selectivity for Na^+ over K^+. To allow the incorporation of CB[6] into a vesicle or lipid bilayer, the periphery of CB[6] was functionalised with alkyl chains.[464] Proton transport across the CB[6]-containing vesicle membrane was evaluated using a pH-sensitive fluorescent dye. These measurements showed that protons were transported through the membrane, but not when the portal of CB[6] was blocked with a guest. Alkali metal transport through the CB[6]-vesicle membrane was also demonstrated and the transport activity of the alkali metals was the exact opposite of their binding affinity: $Li^+ > Cs^+ \sim Rb^+ > K^+ > Na^+$. Conductance measurements in a planar bilayer suggested that the transport across the membrane occurred in a channel mechanism, rather than a carrier mechanisms.

7.3.2. *Towards MRI imaging*

In aqueous environments, CB[6] can absorb neutral gasses including Xe.[47] Hyperpolarised ^{129}Xe is a highly sensitive MRI agent, but it has a tendency to accumulate in lipid tissues and has little affinity for biomolecules since it is a non-polar species. [129]Xe encapsulated in an organic cage is more useful for bioimaging, especially if the cage possesses a targeting group. Cryptophanes have been the cage of choice for this purpose, but they are difficult to produce in the quantities required for bioimaging. CB[6] is cheap, easily synthesised and has a moderate affinity for Xe, which suggests that it may be useful as an MRI contrast agent. Initially, it was thought that CB[6] itself was not soluble enough for biomedical applications, and so water-soluble analogues of CB[6] were evaluated as [129]Xe binders. Kim et al.[465] showed that [129]Xe could be encapsulated in CB*[6] and detected by NMR/MRI. Their studies also revealed that the complex persisted for a longer period of time in the presence of Na^+, presumably because Na^+ acts as a lid. More recently, unfunctionalised CB[6] has been as a cage for [129]Xe. Dmochowski et al.[466] have shown that [129]Xe complexed to unfunctionalised CB[6] experiences a downfield shift in its [129]Xe NMR spectrum, and while it exchanged quickly, it was still slow enough to be observed on the NMR timescale. In 1 μM human serum hyperpolarised [129]Xe@CB[6] could be detected at 1.8 pM. The [129]Xe@CB[6] complex has also been detected *in vivo*, opening the possibility of CB[6] being used as an MRI contrast reagent.[467] Conjugation of a targeting group (such as antibody or aptamer) to CB[6] may allow for targeted imaging. Despite the low solubility of CB[6] in water, biological fluids contain salt molecules that solubilise CB[6], meaning that CB[6]'s poor solubility is not an obstacle for biological applications.

Chapter 8

Applications of Cucurbit[7]uril-Ultrahigh-Affinity Host–Guest Complexes

8.1. Biotin–Streptavidin and Ultrahigh-Affinity Host–Guest Pairs

Biotin–streptavidin (BT–SA) is a natural high-affinity and high-fidelity binding pair ($K_a = 10^{13}$ M^{-1})[468] with a very slow dissociation rate ($k_d = 10^{-6}$ s^{-1}).[469] Such binding characteristics have led to its use across the life sciences, mainly for the immobilisation or purification of tagged bio-molecules. Streptavidin itself is usually immobilised through a surface lysine residue, and BT can be easily functionalised via its carboxylate group which is an ideal location for conjugation to biomolecules since it is not involved in molecular recognition. Despite the obvious advantages, there are limitations to BT–SA system. Since BT is a naturally occurring biomolecule, there is a significant amount of BT and biotinylated proteins in cells. Therefore, false positives are often detected particularly during imaging, which may cause misreading of data. Because SA is a protein (albeit a very stable one), it is susceptible to proteases. Furthermore, because SA is a large (~53 kDa) protein, it cannot be easily internalised into cells or cross cellular compartments, which reduces its utility in bio-assays especially with live cells.

Recovery of proteins attached to SA is also challenging. The BT–SA interaction is very stable, as such harsh denaturing conditions are required to disassociate the interaction and thereby release the proteins. These conditions can be damaging to proteins of interest. What is needed is an artificial, bioorthogonal, stable small molecule pair with strong reversible binding, and excellent fine-tuning potential.

The CB[7]–guest system is an excellent candidate for such an artificial system. The binding affinities of the strongest CB[7]–guest pairs surpass that of BT–SA to form kinetically[470] and thermodynamically[32] stable complexes. CB[7] is small, around 50-fold smaller than the BT–SA complex. The binding affinity between CB[7] and its guest is extremely robust. The guest molecules can be synthetically modified to modulate the binding affinities (from 10^9 to 10^{17} M^{-1}).[30,31] The host–guest interactions are dynamic so that the guests are exchangeable for one with a higher affinity. In addition, the stability of an already formed complex can be modulated by external stimuli. For example, the affinity of ferrocene can be modulated by two orders of magnitude by redox chemistry.[336]

Such merits suggest that CB[7]–guest is a strong candidate to serve as a useful supramolecular tool for various applications. In this chapter, specific applications of the high-affinity host–guest pairs will be showcased in the context of these advantages showing that they are not only a replacement for BT–SA but are, in fact, superior.

8.2. Immobilisation of Biomolecules

The BT–SA pair has long been used to immobilise biotinylated biomolecules onto surfaces as a route to prepare biosensors since a strong interaction is needed to hold the biomolecules in place. After the discovery of CB[7]-based ultrastable synthetic binding pairs in 2005, these synthetic binding pairs were envisioned as a synthetic replacement for BT–SA. The first application was the immobilisation of an enzyme, glucose oxidase (GOx) on a gold surface to prepare a biosensor, reported by Kim *et al.*[69] (Figure 8.1(a)). Multiallyloxylated CB[7] was immobilised onto an alkene-functionalised SAM using a metathesis reaction. GOx conjugated with ferrocenemethylammonium (FA) was immobilised on the CB[7] surface by simple immersion of the CB[7]-anchored surface in a solution of FA–GOx (Figure 8.1(b)). The GOx immobilised on the surface worked as a glucose sensor to detect millimolar concentration of glucose,

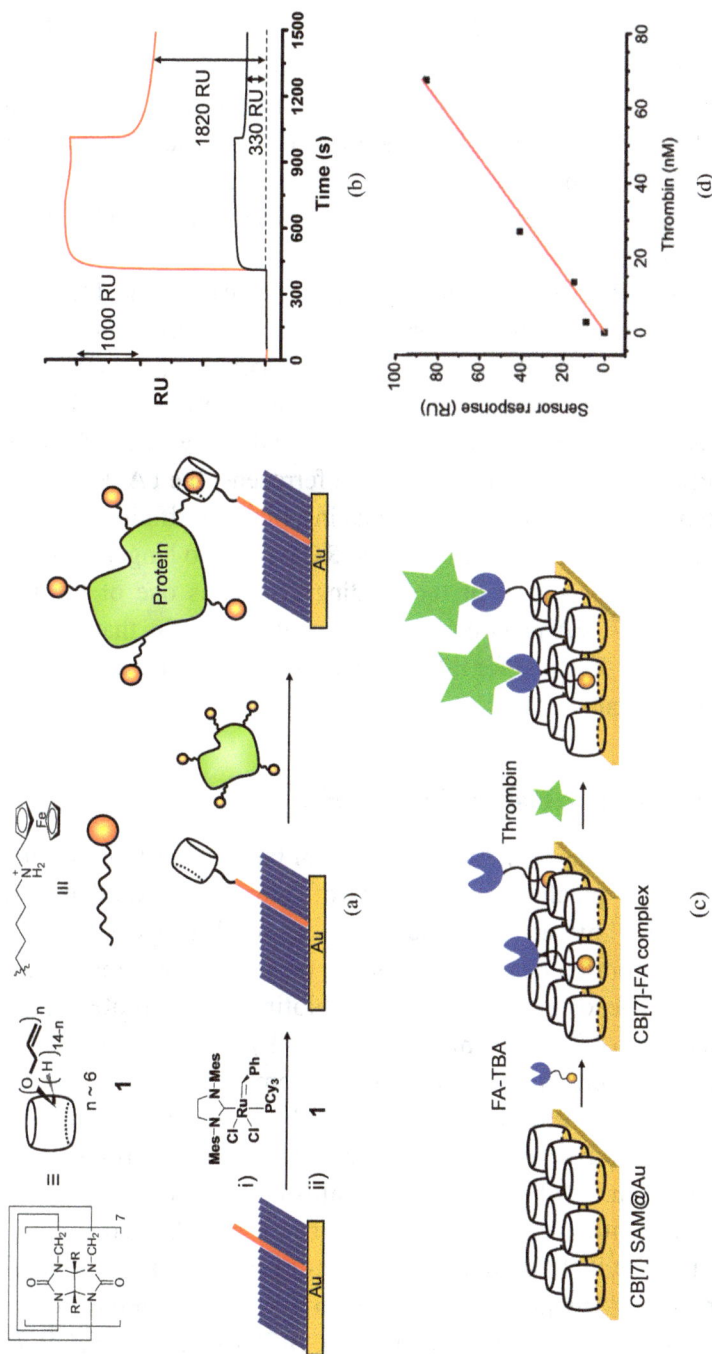

Figure 8.1. (a) Immobilisation of a protein on gold surface using the FA–CB[7] interaction. (b) SPR sensorgram showing immobilisation of FA–GOx (red line) and GOx (black line) on a CB[7]-anchored gold substrate. (Modified from Ref. 69) (c) Preparation of a thrombin biosensor. (d) SPR response to different concentrations of thrombin. (Modified from Ref. 350)

demonstrating a potential use of CB[7]–FA pair as a replacement of BT–SA. Later, it was found that CB[7] itself can form self-assembled layers on gold surfaces as demonstrated by Li *et al.*[76] The surfaces prepared in this manner have also been used to immobilise an FA-conjugated aptamer that can sense thrombin (Figure 8.1(c)).[350] The surface was utilised as a surface plasmon resonance sensor chip, which detected thrombin with sub-ppm sensitivity.

The groups of Brunsveld and Jonkhejim used native chemical ligation to prepare protein conjugates with a single ferrocene moiety, thereby enabling site-specific immobilisation at the C-terminus.[347] An FA-conjugated RGD peptide which is known to promote cell adhesion was immobilised on a gold electrode coated with CB[7] and then cells were cultured on the prepared surface.[348] Upon oxidisation of the ferrocene, the FA–RGD dissociated from the CB[7] surface by reducing binding affinity between CB[7] and FA, and the cells attached to the RGD were no longer adhered. Such electrochemical control of the binding affinity is one of the key advantages of CB[7]–FA pairs over BT–SA, which suggests that CB[7]–FA pairs have great potential for the preparation of electronic devices with biological components.

8.3. Supramolecular Fishing for Proteins

The BT–SA system has been the workhorse of targeted proteomic studies for decades. Typically, biomolecules of interest are labelled with BT and enriched by an SA-functionalised solid support, usually SA-beads. The enrichment is normally performed by incubating SA-immobilised beads in a solution of the biotin-labelled proteins, and then the proteins are eluted from the SA-beads by breaking the BT–SA interaction. However, this system suffers from intrinsic drawbacks, mainly the unwanted enrichment of naturally occurring biotinylated proteins which manifest as false positives and reduces the efficiency of protein enrichment by blocking binding sites on SA. The other disadvantage is that eluting the target proteins from the SA-bead requires harsh conditions, which can damage proteins of interest. To solve these issues, Kim *et al.*[84] replaced SA-bead with a synthetic alternative, the CB[7]-bead. Multihydroxyl-functionalised CB[7] (multihydroxyCB[7]) was conjugated to *N*-hydroxysuccinimide-functionalised sepharose beads to provide CB[7]-beads that can capture FA-labelled proteins using the host–guest interaction between CB[7] and FA (Figure 8.2). Specifically, they selectively labelled the plasma membrane proteins on the cell surface with FA succinimide. After the cells were lysed, the FA-labelled proteins were captured on CB[7]-beads by incubation of the bead with the cell lysate. The labelled proteins were released from the beads by treatment with a higher affinity guest, 1,1′-bis(trimethylammoniummethyl)ferrocene (BAFc), under ambient conditions; the higher affinity guest efficiently displaced the FA-labelled proteins captured on CB[7]-beads. The proteins enriched were, as expected, membrane proteins, exclusively, and the ambient protein recovery was just as effective as the harsh conditions used with BT–SA. These results suggest that CB[7]–FA can not only replace BT–SA but also provide even more efficient protein enrichment with easier processability.

Later, Kim *et al.*[471] extended this concept to a specific cytosolic protein by use of a drug molecule as a "molecular bait" to fish out a target protein from the cytosol. Suberanilohydroxamic acid (SAHA) was selected as the drug molecule which can recognise histone deacetylase (HDAC) in cytosol. They conjugated adamantylamine (Ad) to the drug by a bridging benzophenone moiety to provide a photocrosslinkable probe for HDAC (Ad–SAHA). The Ad and benzophenone moieties were used for ultrastable binding to CB[7] and photo-crosslinking of Ad to the target proteins after the drug recognises the target protein, respectively. Incubating Ad–SAHA with cells permeabilised with a small amount of detergent followed by UV irradiation allowed the transfer Ad to HDAC in a live cell; HDAC was successfully enriched by CB[7]-beads after cell lysis. This work demonstrated that the CB[7]-bead system could be utilised to enrich a specific cytosolic protein by utilising a drug as a molecular bait. In principle, this strategy can be extended to other drug molecules for other important target proteins and may bridge the gap between supramolecular chemistry and chemical biology.

Using high-affinity pairs is not the only supramolecular strategy for protein enrichment. Proteins with an N-terminal phenylalanine can also be enriched. Urbach and Isaacs[472] demonstrated this by attaching CB[7] to a support sepharose bead: they first prepared mono-azide-functionalised

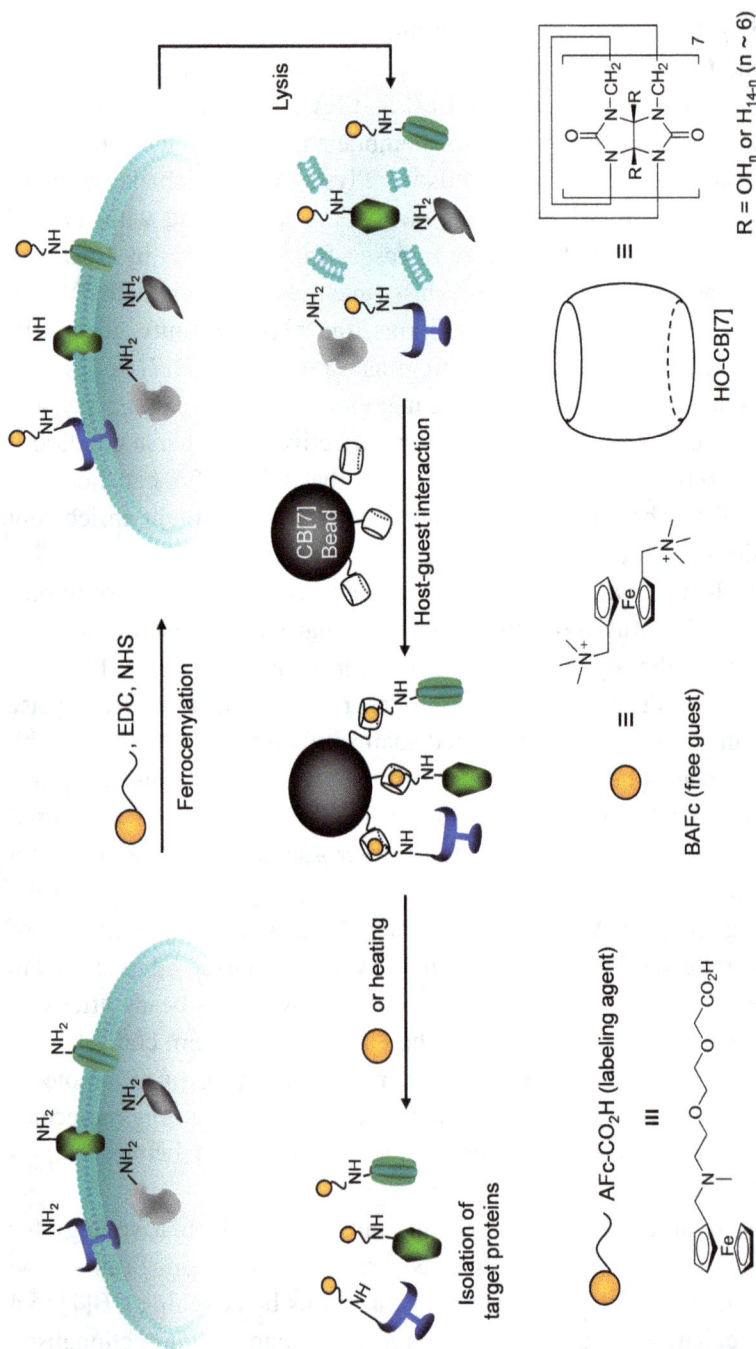

Figure 8.2. Strategy for isolation of plasma membrane proteins using an ultrastable synthetic binding pair system. Lysine residues on the cell surface were ferrocenylated. The ferrocenylated proteins are selectively captured by CB[7]-beads. (From Ref. 84)

CB[7] and attached to it an alkyne-functionalised bead via a Cu⁺-catalysed click reaction. This bead could capture proteins with an N-terminal phenylalanine, notably human insulin. The proteins were captured by the CB[7]-bead, and the non-specifically adsorbed proteins were washed away by washing with buffer. The target proteins were then released by addition of a competitive guest. Although there was a definite preference for proteins with an N-terminal phenylalanine, proteomic analysis revealed that the beads enriched other proteins that did not have the N-terminal epitope. Since their bead is conjugated to the solid support via azide–alkyne click chemistry, it seems likely that their bead is more robust than the ester-linked bead described here. However, their strategy targets N-terminal phenylalanine, which may be less effective in terms of binding affinity compared to that of FA or Ad. Furthermore, the presence of extra hydroxyl groups in multihydroxyCB[7] might help reduce non-specific interactions with proteins, which is another point to consider for further exploration in this direction. In the future, a combination of the synthetic and strategic approaches may be the best way forward.

8.4. Host–Guest FRET Pairs for Contents Mixings Assays

Förster Resonance Energy Transfer (FRET) is a widely used technique to investigate binding or other proximal interactions in biology. FRET pairs are well suited to this purpose because a FRET signal is only observed when two complimentary fluorophores (a FRET donor and acceptor) are close to each other. FRET pairs inside vesicles have been used to detect vesicle fusion, when two vesicles fuse and the contents mix, which causes the FRET pair to dissociate or associate with a decrease or increase in fluorescence, respectively. For example, DNA-hairpin molecular beacons are one tool that uses FRET. In the initial state, FRET is observed since the FRET dyes are at the termini of the hairpin (one at each end). Upon fusion with another vesicle containing a complementary strand of DNA, the hairpin unravels and a double helix forms, which moves the dyes to the opposite end of the DNA strand, thereby abolishing FRET. Another approach is to use self-quenching dyes (such as sulphorhodamine B), which upon fusion with an empty vesicle switches on fluorescence by virtue of a dilution effect. Both of these assays are useful but have a low signal-to-noise ratio, poor resolution and slow kinetics, so they cannot unravel intricate multistep processes.

Kim *et al.*[103] reported a small, ultrastable synthetic host–guest interaction-based FRET pair and used it to study content mixing upon protein-mediated vesicle fusion (Figure 8.3). CB[7] and adamantylamine (Ad) were functionalised with Cy3 and Cy5 as FRET donor (CB[7]–Cy3) and

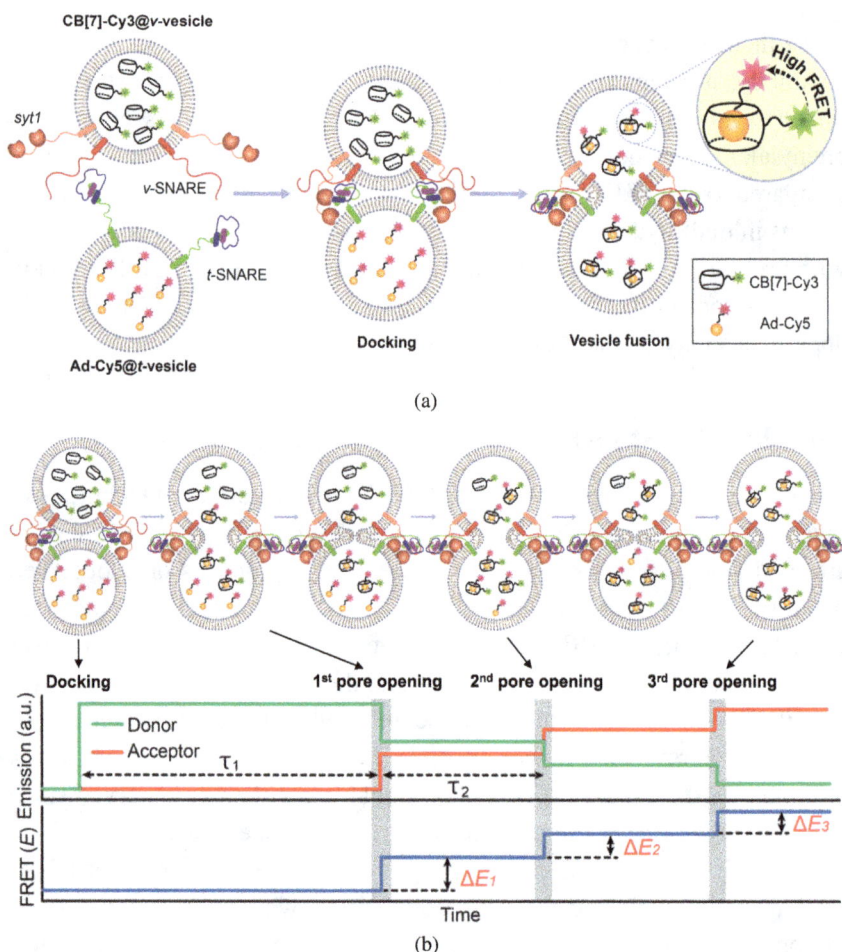

Figure 8.3.   (a) Schematic illustration of SNARE-mediated vesicle fusion using the supramolecular FRET pair. (b) Step-by-step mechanism with FRET evidence of pore opening events; the blue line corresponds to FRET. (Modified from Ref. 103)

acceptor (Ad–Cy5), respectively; when they form a host–guest complex, FRET occurred. This pair was used to investigate single vesicle fusion. In order to make two vesicles fuse, soluble-*N*-ethylmaleimide-sensitive factor attachment protein (SNARE) receptors were incorporated into their membrane, since these proteins are known to mediate membrane fusion in nature. Two kinds of vesicles were prepared: one with CB[7]–Cy3 inside, and *v*-SNARE reconstituted into its membrane, termed the donor vesicle; the other had Ad–Cy5 inside, and *t*-SNARE reconstituted into its membrane, termed the acceptor vesicle. As the two vesicles docked and the contents were mixed, a FRET signal emerged in flickering bursts. These flickering bursts suggested that the vesicles had docked and formed pores which opened and closed several times; these events were not observed using self-quenching dyes or molecular beacons. The work described here demonstrates the power of using a small, synthetic system to study a complex biochemical phenomenon. It is anticipated that such supramolecular FRET pairs may be useful in other biological systems where sensitive, high-resolution, bioorthogonal probes are required; for example, the fusion of organelles during cellular processes.

8.5. Supramolecular Velcro for Underwater Adhesion

Synthetic underwater adhesives are a growing research area in materials science because of their potential as surgical adhesives for biomedical applications. Most efforts to prepare underwater adhesives follow biomimetic approaches, such as modifying adhesive proteins that are produced by marine organisms.[473] However, these approaches have yet to produce strong underwater adhesives, and their adhesion is not easily modulated or reversed by external stimuli. Kim *et al.*[95] reasoned that the CB[7]–FA pair should be ideally suited for use in underwater adhesive applications since these pairs form ultrahigh-affinity complexes in water, and the interaction can be modulated with redox chemistry. (Figure 8.4).

The approach demonstrated here is reminiscent of the macroscopic material, Velcro®. Velcro works on a macroscale by fastening hooks into loops. In this case, a similar mechanism was used but on a molecular level. The loop was made from mono-allyoxylated CB[7], and the hook comprised of a cationic ferrocene derivative. CB[7] and FA were each grafted

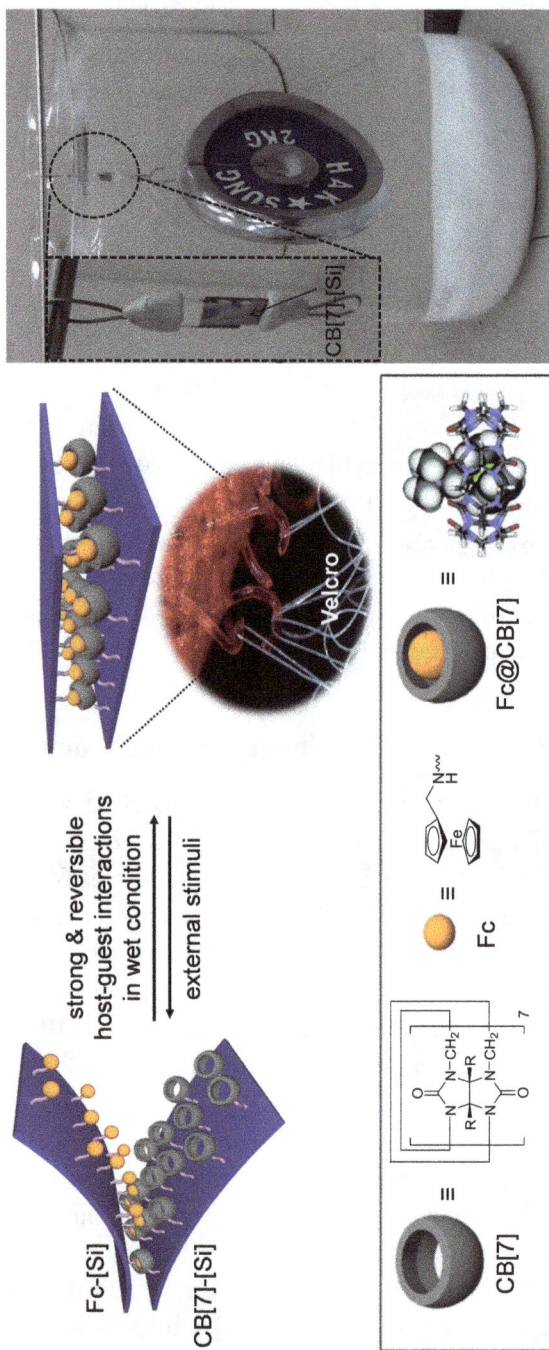

Figure 8.4.   (a) Supramolecular Velcro for underwater adhesion and (b) load-bearing experiments in water. (From Ref. 95)

to a polymer-coated solid surface. When the surfaces (1×1 cm^2 area) were wet and pressed together, they could hold a weight of 2 kg, but when they were air dried first, they held double the weight. The surfaces adhere to each other with strong adhesion forces (1.12 MPa); stronger than commercial Velcro (0.08 MPa) and double-sided tape (0.74 MPa). The surfaces could either be separated by mechanical force or by oxidation of FA with hypochlorite solution. Reduction with ascorbic acid caused the adhesion property to return, albeit with reduced performance. One area that needs to be improved is the recovery of performance after an oxidation–reduction cycle; the drop in performance is probably due to the ferrocenium species being quenched, chemically. This work represents the transformation of a molecular recognition event — an ultrahigh-affinity host–guest pair — into a macroscopic property. Other supramolecular adhesives, such as those based on the cyclodextrin–ferrocene interactions,[474] can only hold a fraction of the load of the CB[7] material, further underscoring the remarkable properties of the CB[7]–guest pairs.

8.6. Activation of Therapeutic Nanoparticles

Transferring supramolecular chemistry into live cells is a formidable challenge, but a potentially rewarding one to overcome since it may allow bioorthogonal control over synthetic systems *in vivo*. A particularly attractive line of research is to use supramolecular interactions for activation of therapeutic effects in a spatiotemporally specific manner, which may help mitigate off-target toxic effects. With this in mind, Rotello and Isaacs[36] have developed recognition-mediated activation of therapeutic gold nanoparticles (AuNP) (Figure 8.5(a)). The recognition-mediated activation utilised a controllable interaction between CB[7] and 1,6-diaminohexane (DAH) which can be replaced with a stronger guest which has much higher binding affinity, namely adamantylamine (Ad) by a simple treatment. They prepared CB[7]-capped DAH-terminated gold nanoparticles, which after cellular uptake, were not cytotoxic since they cannot escape the endosome. However, in response to treatment of Ad, the NPs rapidly escaped from endosomes to become cytotoxic, since Ad formed a strong complex with CB[7] and exposed the amine moieties on the NPs (Figure 8.5(b)). These results demonstrated that a guest exchange

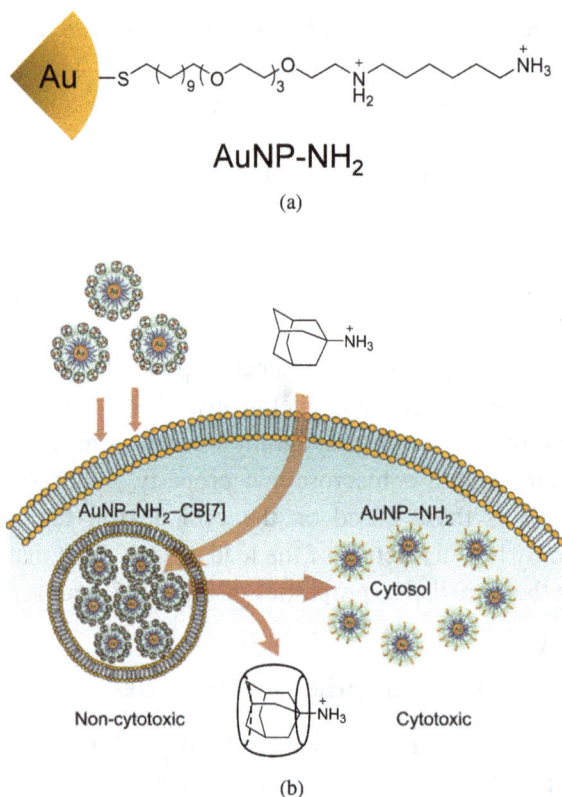

Figure 8.5.   (a) Structure of cytotoxic nanoparticles and (b) cytotoxicity activation mechanism. (Modified from Ref. 36)

in live cells allows for delicate control of the cytotoxicity of NPs. In a follow-up work, the same group applied the same principles to regulate the exocytosis of NPs.[475] These works are significant in that it shows the great potential of CB[7]-based controllable host–guest chemistry as a novel method for controlling the intracellular behaviour of NP and enhancing localised therapeutic effects of the cytotoxic NP at the target sites.

8.7. Regulation of Catalysis

In nature, protein–protein or protein–ligand interactions play important roles in regulating the functions of proteins. With this in mind, CB[7]

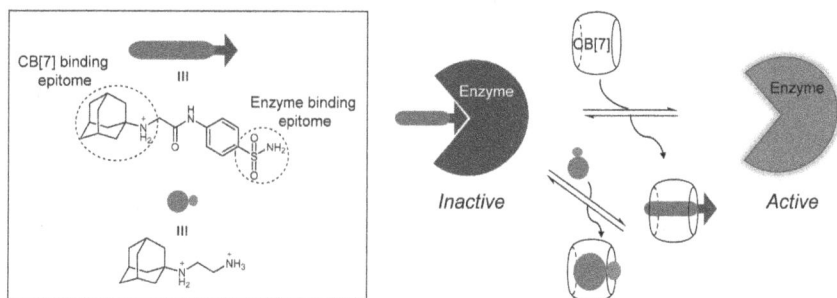

Figure 8.6. Control of BCA activity using a two-faced molecule and CB[7]. (From Ref. 476)

host–guest pairs have been used as bioorthogonal switches to regulate protein activity or chemical catalysis in a biological context. Isaacs *et al.*[476] have demonstrated control of biological catalysis using CB[7]. They prepared a supramolecular switch by synthesising a bifunctional molecule (Figure 8.6). One face of the molecule was designed to bind to CB[7] and the other to bovine carbonic anhydrase (BCA). When the molecule is bound to BCA, it inhibits the activity of the enzyme. Once CB[7] is added, the inhibitor becomes too bulky and dissociates from the BCA, restoring the activity of the enzyme. The process can be reversed by addition of a high-affinity guest that causes the release of the two-faced molecule from CB[7], which in turn, reinhibits the BCA. The key requirement for this strategy is that the enzyme has a sterically demanding active site; otherwise, the addition of CB[7] does not cause dissociation of the inhibitor. Nevertheless, this work demonstrated a CB[7]-based switch that allowed control over protein activity.

CB[7]-capped nanoparticles can be used to regulate the activity chemical catalyst inside living cells. Rotello *et al.*[105] demonstrated supramolecular regulation of catalysis inside cells (Figure 8.7). They used AuNP, the surface of which was capped with CB[7] that forms a host–guest complex with a dimethylbenzylammonium on the NPs. The SAM layer was impregnated with a hydrophobic organometallic catalyst which could not escape because it was held by hydrophobic interactions between the ligands of the catalyst and the alkyl chains in the SAM. Because of the presence of CB[7] on the surface of the NP, small molecules cannot reach into the catalyst either. However, the addition of Ad caused the removal of

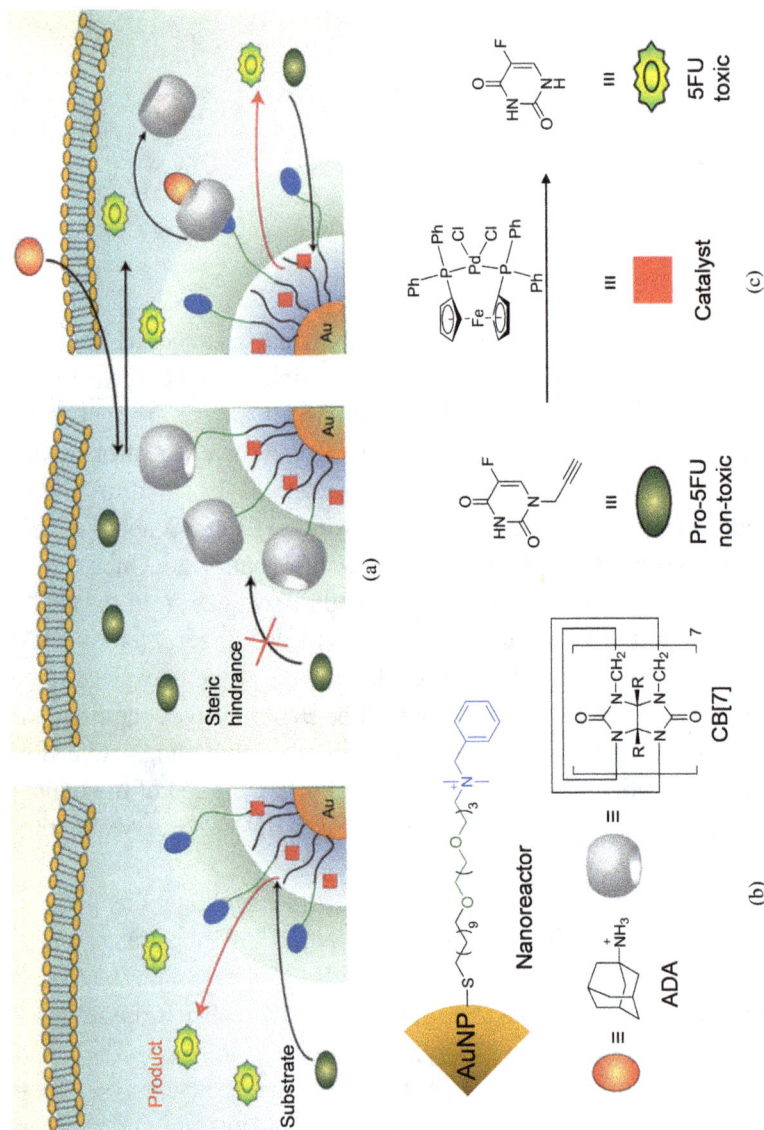

Figure 8.7. (a) Nanogated release of catalysts from CB[7]-capped nanoparticles by guest exchange. (b) Structure of nanoparticles. (c) Prodrug activation mechanism. (Modified from Ref. 105)

the CB[7] from the NP upon forming an extremely stable host–guest complex with CB[7] and, consequently, made small molecules accessible to the catalyst in the NP. They demonstrated this principle with a cleavage reaction of allylcarbamate in live cells and applied it for conversion of a prodrug to the active drug in a live cell. Cell viability assays showed that addition of Ad to the cells treated with the prodrug and the catalyst-NP had a cytotoxic effect. This work is another good example of how CB[7] host–guest chemistry and guest exchange is robust and specific enough to work in a cellular context. More generally, this work demonstrates the potential of CB[7]-based supramolecular chemistry for controlling the activity of an organometallic catalyst in a nanoreactor in live cells, without severe undesired cytotoxicity. This work may be extended to develop a new platform for bioimaging and therapeutic applications.

8.8. Summary

The ultrahigh-affinity CB[7]–guest pairs have come a long way since their affinities were reported in 2005. The key features that set them apart from other host molecules are their ultrahigh-affinity binding and guest discrimination. β-cyclodextrin, a similarly-sized macrocycle, rarely forms host–guest complexes with affinities greater than 10^6 M^{-1}, and there is negligible discrimination based on charge. Most biological pairs, such as BT–SA or antibody–antigen, are highly specific for each other and have evolved to recognise to each other with high fidelity. In contrast, CB[7] is a synthetic host molecule for several different ultrahigh-affinity guests, and the binding affinity of these guests can be tuned in the range of 10^9–10^{17} M^{-1}. Other advantages of the ultrahigh-affinity complexes include robust host–guest complexes even in salt-rich or biological environments, which is impressive considering that the presence of metal ions, such as those in physiological conditions, is known to weaken host–guest complexes. In addition, the binding affinity can be controlled by other guest molecules or by external stimuli. The facile attachment and detachment is reminiscent of a mechanical "latch"; as such, this system has been termed a "supramolecular latching system".[477]

The merits listed above are critical for CB[7]'s current and future biological applications. The ultrahigh-affinity binding in biological environments with little interference from biomolecules means that they can be considered bioorthogonal binding tools. The bioorthogonal ultra-high-affinity guest pairs have been used to immobilise or capture labelled proteins. The well-defined guest exchange behaviour is transferable to live cells, where it was used to open the gate to a nanoreactor inside living cells. The small size of CB[7]–guest pairs relative to that of biomolecules, such as proteins, means that the synthetic binding pair can be used not only as bioorthogonal probes but also as high-resolution imaging agents. The example highlighted in this chapter described a supramolecular FRET pair which unravelled the discrete steps involved in a SNARE-mediated vesicle fusion. In future, it is anticipated that these pairs may be used as cellular probes to elucidate the mechanisms of complex biological processes.

Chapter 9

Perspectives: Outstanding Challenges and Opportunities

In this book, we have discussed the merits of cucurbiturils and shown how such merits can be exploited, with great success, in the fields of chemistry, materials science and biology. The CB family shows remarkable binding selectivity and affinity, which distinguishes them from other host molecules. Nevertheless, many challenges and opportunities remain to fully capitalise on the unique features of the CB family. Our thoughts outlining what these challenges are and how they might be met are discussed in this chapter.

9.1. Scalable Synthesis of CBs, Their Derivatives and Analogues

Much effort has been devoted to improving the synthesis, separation and purification of each of the CB[n] family members, in particular CB[7] and CB[8]. While they can be purchased from chemical suppliers, they are still much more expensive than the respective cyclodextrins. What is required is a high yielding, scalable synthesis of the higher family members ($n > 6$). Templating agents and effects have been investigated to this end, with limited success so far; however, further investigation is merited. Methods need to be developed for reactions on a larger scale (i.e. >1 kg). Another consideration for large-scale synthesis is that it generates a lot of highly acidic waste. A procedure to mitigate this waste, for example by recycling, would be helpful for industrial scale synthesis. A robust separation and purification procedure that affords the desired family member in high purity is also required.

While some progress has been made in the synthesis of CB[n] and derivatives, scalable methods to synthesise functionalised CB[n] are more urgently required. There are three oxidative methods to prepare hydroxylated CB[n]. All of these methods share a common problem: the extent of oxidation is difficult to control and separating the products is quite difficult and reduces the yield. An alternative approach is the building block synthesis, where modified glycoluril surrogates are used to form CB derivatives (both these strategies were discussed in Chapter 2). Neither of these strategies are yet able to produce functionalised CBs in large amounts and good yields. The problem is particularly apparent with CB[8].

Unlike cyclodextrins, achiral cucurbituril molecules cannot discriminate between enantiomers. In the future, chiral derivatives or analogues of CB[n] may emerge that will enable enantioselective catalysis and separations. However, there are currently few chiral derivatives or analogues of CBs that can discriminate between enantiomers well. For optimal chiral recognition, the portal of the CB[n] may need to be desymmetrised, but not in a way that will significantly compromise the recognition properties. In some ways, chiral cucurbiturils are at the same stage of development as achiral CBs were pre-2000; what is needed is an enabling synthetic method to access chiral CB derivatives and analogues.

New analogues that have unexpected properties may also be discovered. Inspiration can be taken from the bambusurils, which are less than a decade old but have emerged as very effective anion binders. The hemicucurbiturils are a family that is experiencing growth at the moment; they are less rigid than CBs so can be rendered chiral more easily. Chiral hemicucurbit[6]uril[152] and hemicucurbit[8]uril[478] have been prepared recently, although these have not yet been used to discriminate between chiral guests.

Effective synthetic methods to access cucurbiturils, their derivatives and analogues are crucial to make CBs and CB-related compounds and commodities for materials scientists and engineers.

9.2. Kinetics and Mechanistic Studies of Host–Guest Inclusion

The thermodynamics of host–guest binding have been investigated extensively, and many of the underlying processes are now reasonably well understood. On the other hand, there are far fewer studies on the kinetics of host–guest binding and those of guest exchange. Such studies are crucial for gaining mechanistic insights into how host–guest complexes form and dissociate. Such considerations are important not only for a fundamental understanding of CB host–guest chemistry but also for advanced applications such as molecular machines and switches.

The mechanism of CB[6]-guest binding has been investigated with wide aromatic guests that have a slow association rate. These experiments revealed that binding with these guests occurs in two steps. First, an exclusion complex forms between the cation and the CB[6] portal, and second, a "flip-flop" mechanism occurs where the hydrophobic moiety inserts into the cavity.[53,210] In the case of CB[7], the evidence suggests that for small CB[7] guests such as berberine[213] and methylviolgen[212] a single-step mechanism dominates. For many guests, including important guests such as ferrocene and adamantane, the binding mechanism is not well understood. Moreover, there is little information about how guest exchange reactions may occur. Supramolecular FRET pairs may be able to provide some insight into this mechanism since they can report association and dissociation events with high sensitivity.

9.3. Potential Applications as Molecular Containers

Supramolecular host molecules have long been envisioned as artificial enzyme mimics. CBs have been employed as reaction containers that facilitate and influence the stereochemical course of the reaction. A persistent problem with this approach is product inhibition, in some cases the product cannot be removed from the CB at all. There have been few examples of truly CB-catalysed reactions. A recent example has suggested that product inhibition can be overcome with certain substrates.[112] More general approaches to tackle this issue are required. As mentioned above, chiral derivatives and analogues of CBs are needed for asymmetric catalysis and separation.

CBs have great potential as sensors. Among the sensors developed, functionalised surfaces have been established as a versatile and powerful platform. A particularly powerful combination is CB-functionalised organic field-effect transistors (OFET) which allows detection of picomolar amounts of analytes.[272,274] Other high-sensitivity CB-based sensors may emerge in the future. CB[8]-based sensors that detect the binding of a second guest may have a bright future because the selectivity and desired properties could be tuned by the auxiliary guest. Such sensors may become marketable products.

CBs are a promising candidate for components in molecular machines since they have well-defined and dynamic host–guest chemistry and CBs themselves are inert to photochemical and electrochemical stimuli. Basic components such as switches, shuttles and molecular loop locks have been demonstrated; however, this aspect of CB chemistry is still in its infancy compared to cyclodextrins. Cyclodextrins can both be covalently and non-covalently modified in the same system, which allows the creation of mechanical devices such as molecular motors, ratchets and actuators.[479]

9.4. Applications in Materials Science

Supramolecular materials are another area where CB[n] have not yet reached their potential. Porous CB has been demonstrated as a gas sorption material, a proton conductor and a lithium-ion conductor. Since CBs are thermally and chemically stable and made from cheap starting materials (relative to other porous materials), they may be promising materials for storage, controlled release, separation and removal of molecules of interest.

There have been some notable examples of supramolecular materials. Among these, proof-of-concept works have shown that CB[n] hydrogels are a promising soft material with biomedical applications. They have impressive ability to act as an extracellular matrix to grow therapeutic cells even inside a live animal. Further development in this direction may lead to real therapeutic treatments, such as long-term release of therapeutic proteins and tissue engineering.

CB-based materials such as 2D polymers and nanocapsules appear to have a bright future as a tailorable material; these materials are quite rare because they can be post-synthetically modified. To date, the 2D polymer has been used as a permselective membrane, but it may also be used as selective transport material for other purposes such as desalination. The nanocapsules have been used as both drug delivery and an imaging agent; in the future, they may be used as a multimodal theranostic platform. Another exciting development is the preparation of supramolecular organic frameworks (SOFs) using CB[8] to stabilise methyl viologen radical dimer; these materials may have use as selective transport, storage or release materials. CB[8] chemistry has been combined with microfluidics in order to create microscale supramolecular architectures; in the future, these materials may be used biomedically.[400]

The ability of CB[*n*] to form a nanogap between metallic surfaces has been used to induce plasmonic enhancements. So far, this has been applied mostly towards single and few molecules SERS measurements since the CB[*n*] cavity can securely localise guests in the nanogap. The plasmonic enhancement effects may also be applied to other areas of photonics such as photovoltaic devices.

9.5. Applications in Biology

CB[*n*] have been used in biological recognition and several high-affinity peptide epitopes have been identified. These epitopes have been used to modify protein non-covalently and artificially induce protein dimerisation. Since CB[8], in particular, recognises natural epitopes, they can easily be incorporated into proteins, which may expedite their transition into useful applications, for example, preparation of functional protein assemblies.

CB[*n*] have been used in drug delivery where they have been shown to stabilise and solubilise drug molecules. More recently, hierarchical CB-based materials have been developed which allow a greater number of drug molecules to be delivered. Future developments may include smart theranostic materials that target particular diseases.

In Chapter 8, high-affinity host–guest pairs were discussed in the context of biotin–streptavidin (BT-SA), and it was shown the CB[7]–guest pairs might supplant BT-SA in several biological applications. These guest pairs can be environed for other uses. CB[7]–dyes may be used as high-resolution imaging agents of guest labelled biomolecules since CB[7]–dyes can enter live cells (unlike protein-based imaging agents), and its host–guest chemistry is bioorthogonal and robust in physiological environments. CB[7] pairs have also been used for protein enrichment, and the advantages of this method compared to BT–SA were highlighted. Enrichment experiments provide a lot of biological information; the emergence of an orthogonal pair means that both could be used in the same experiment.[477] To illustrate here, double labelling and enrichment allows two questions to be asked at the same time. For example, two organelle-specific labelling methods could be applied and proteins that are labelled and enriched by both methods suggest that they are involved in trafficking between those organelles.

In a more recent development, CB[6] was used to label biomolecules as shown by Francis *et al.*[110] They took CB[6]'s ability to perform azide–alkyne cycloaddition reactions and applied it to suitably labelled biomolecules. In principle, this approach can be applied as widely as conventional click reaction, with the advantage of it being copper-free.

9.6. Further Reading

In this book, the broad field of cucurbiturils chemistry has been surveyed. It is not possible to do justice to all the interesting chemistry, supramolecular chemistry and applications that have been achieved so far. Shown in the table below (Table 9.1) are a number of review articles written from various perspectives that readers are referred to for a more detailed look at areas that interest them.

Table 9.1.   Reviews covering various aspects of cucurbiturils chemistry.

Review title (author, year)	Refs.
Cucurbituril (Mock, 1995)	480
Self-assembly of interlocked structures with cucurbituril, metal ions and metal complexes (Kim, 1998)	481
Mechanically interlocked molecules incorporating cucurbiturils and their assemblies (Kim, 2002)	17

(Continued)

Table 9.1. (*Continued*)

Review title (author, year)	Refs.
Rotaxane dendrimers (Kim, 2003)	482
Cucurbituril homologues and derivatives: New opportunities in supramolecular chemistry (Kim, 2003)	21
The cucurbit[*n*]uril family (Isaacs, 2005)	483
Functionalised cucurbiturils and their applications (Kim, 2007)	451
Supramolecular assemblies built with host–stabilised charge–transfer interactions (Kim, 2007)	311
Cucurbituril encapsulation of fluorescent dyes (Nau, 2007)	484
Cucurbit[*n*]urils: From mechanism to structure and function (Isaacs, 2009)	126
The potential of cucurbit[*n*]uril in drug delivery (Wheate, 2011)	485
The mechanism of cucurbiturils formation (Isaacs, 2011)	125
Deep inside cucurbiturils: Physical properties and their inner cavity (Nau, 2011)	133
Molecular recognition of amino acids, peptides and proteins by cucurbit[*n*]uril receptors (Urbach, 2011)	419
Gas-phase cucurbit[*n*]uril chemistry (Dearden, 2011)	215
Cucurbituril complexes of redox-photoactive guests (Kaifer, 2011)	486
Cucurbiturils as versatile receptors for redox-active substrates (Kaifer, 2011)	487
Ultrastable host–guest complexes and their applications (Kim, 2011)	488
Applicable properties of cucurbiturils (Keinan, 2011)	489
Supramolecular assemblies constructed by cucurbiturils-catalyzed click reaction (Tuncel, 2011)	229
Cucurbituril: At the interface of small molecule host–guest chemistry and dynamic aggregates (Scherman, 2011)	490
Encapsulation of drug molecules by cucurbiturils: Effects on their chemical properties in aqueous solution (Macartney, 2011)	253
Molecular selective binding and nanofabrication of cucurbituril/cyclodextrin pairs (Liu, 2011)	491
From small cucurbituril complexes to large ordered networks (Buschmann, 2011)	492
Novel supramolecular hosts based on linear and cyclic oligomers of glycoluril (Sindelar, 2011)	493

(*Continued*)

Table 9.1. (*Continued*)

Review title (author, year)	Refs.
Fluorescence response of alkaloids and DAPI on inclusion in cucurbit[7]uril: Utilization for the study of the encapsulation of ionic liquid cations (Biczók, 2011)	494
Supramolecular assemblies of thioflavin T with cucurbiturils: prospects of cooperative and competitive metal–ion binding (Mohanty, 2011)	495
Cucurbituril chemistry: A tale of supramolecular success (Masson, 2012)	7
Cucurbit[n]uril-based coordination chemistry: From simple coordination complexes to novel poly-dimensional coordination polymers (Tao, 2013)	339
Cucurbituril: A promising organic building block for the design of coordination compounds and beyond (Cao, 2013)	496
Stimuli-responsive systems constructed using cucurbit[n]uril-type molecular containers (Isaacs, 2014)	325
The hydrophobic effect revisited-studies with supramolecular complexes imply High-energy water as a non-covalent driving force (Biedermann, Nau and Schneider, 2014)	206
Toward reversible control of cucurbit[n]uril complexes (Kaifer, 2014)	336
Cucurbiturils: Chiral applications (Ghanem, 2014)	497
Dynamic analyte-responsive macrocyclic host–fluorophore systems (Nau, 2014)	25
Cucurbituril-based supramolecular engineered nanostructured material (Tuncel, 2015)	498
Cucurbiturils: From synthesis to high-affinity binding and catalysis (Nau, 2015)	132
Can we beat the biotin–avidin pair?: Cucurbit[7]uril-based ultrahigh affinity host–guest complexes and their applications (Kim, 2015)	499
Cucurbituril-based molecular recognition (Scherman, 2015)	367
Synthesis and separation of cucurbit[n]urils and their derivatives (Tao, Lindoy and Wei, 2016)	118
Cooperative capture synthesis: Yet another playground for copper-free click chemistry (Stoddart, 2016)	306
Ultrastable artificial binding pairs as a supramolecular latching system: A next generation chemical tools for proteomics (Kim, 2017)	477
Synthetic mimics of biotin/(strept)avidin (Smith and Isaacs, 2017)	500
Construction of protein assemblies by host–guest interactions with cucurbiturils (Hou and Liu, 2017)	501

(*Continued*)

Table 9.1. (*Continued*)

Review title (author, year)	Refs.
Cucurbit[*n*]uril-based microcapsules self-assembled with microfluidic droplets: A versatile for supramolecular architectures and materials (Scherman, 2017)	400
Electrochemical properties of cucurbit[7]uril complexes of ferrocenyl derivatives (Kaifer, 2017)	502
The aqueous supramolecular chemistry of cucurbit[*n*]urils, pillar[*n*]arenes and deep-cavity cavitands (Kim, Ogoshi and Gibb, 2017)	503

References

1. Lehn, J.-M. (1995). *Supramolecular Chemistry* (VCH, New York).
2. Atwood, J. L. (1996). *Comprehensive Supramolecular Chemistry*, 1st Edn. (Pergamon, Oxford).
3. Sauvage, J.-P. and Gaspard, P. (2010). *From Non-Covalent Assemblies to Molecular Machines* (Wiley-VCH, Verlag).
4. Bruns, C. J. and Stoddart, J. F. (2017). *The Nature of the Mechanical Bond: From Molecules to Machines* (Wiley-VCH, Verlag).
5. Feringa, B. L. and Browne, W. R. (2011). *Molecular Switches* (Wiley-VCH, Verlag).
6. Behrend, R., Meyer, E., Rusche, F. and Ueber, I. (1905). Condensationsproducte Aus Glycoluril Und Formaldehyde. *Justus Liebigs Ann. Chem.*, 339, pp. 1–37.
7. Masson, E., Ling, X., Joseph, R., Kyeremeh-Mensah, L. and Lu, X. (2012). Cucurbituril Chemistry: A Tale of Supramolecular Success. *RSC Adv.*, 2, pp. 1213–1247.
8. Freeman, W. A., Mock, W. L. and Shih, N. Y. (1981). Cucurbituril. *J. Am. Chem. Soc.*, 103, pp. 7367–7368.
9. Mock, W. L. and Shih, N. Y. (1983). Host–Guest Binding Capacity of Cucurbituril. *J. Org. Chem.*, 48, pp. 3618–3619.
10. Mock, W. L., Irra, T. A., Wepsiec, J. P. and Adhya, M. (1989). Catalysis by Cucurbituril. The Significance of Bound-Substrate Destabilization for Induced Triazole Formation. *J. Org. Chem.*, 54, pp. 5302–5308.
11. Mock, W. L. and Pierpont, J. (1990). A Cucurbituril-Based Molecular Switch. *J. Chem. Soc.*, 21, pp. 1509–1511.

12. Buschmann, H.-J. and Schollmeyer, E. (1997). Cucurbituril and Cyclodextrin as Hosts for the Complexation of Organic Dyes. *J. Incl. Phenom. Mol. Recognit. Chem.*, 29, pp. 167–174.

13. Buschmann, H.-J. and Wolff, T. (1999). Fluorescence of 1-Anilinonaphthalene-8-Sulfonate in Solid Macrocyclic Environments. *J. Photochem. Photobiol. A Chem.*, 121, pp. 99–103.

14. Buschmann, H. J., Jansen, K. and Schollmeyer, E. (1992). Cucurbit[6]uril as Ligand for the Complexation of Lanthanide Cations in Aqueous Solution. *Inorg. Chim. Acta*, 193, pp. 93–97.

15. Buschmann, H.-J., Jansen, K. and Schollmeyer, E. (1998). The Formation of Cucurbituril Complexes with Amino Acids and Amino Alcohols in Aqueous Formic Acid Studied by Calorimetric Titrations. *Thermochim. Acta*, 317, pp. 95–98.

16. Jeon, Y. M., Kim, J., Whang, D. and Kim, K. (1996). Molecular Container Assembly Capable of Controlling, Binding and Release of Its Guest Molecules: Reversible Encapsulation of Organic Molecules in Sodium Ion Complexed Cucurbituril. *J. Am. Chem. Soc.*, 118, pp. 9790–9791.

17. Kim, K. (2002). Mechanically Interlocked Molecules Incorporating Cucurbituril and Their Supramolecular Assemblies. *Chem. Soc. Rev.*, 31, pp. 96–107.

18. Whang, D., Park, K. M., Heo, J., Ashton, P. and Kim, K. (1998). Molecular Necklace: Quantitative Self-Assembly of a Cyclic Oligorotaxane from Nine Molecules. *J. Am. Chem. Soc.*, 120, pp. 4899–4900.

19. Kim, J., Jung, I. S., Kim, S. Y., Lee, E., Kang, J. K., Sakamoto, S., Yamaguchi, K. and Kim, K. (2000). New Cucurbituril Homologues: Syntheses, Isolation, Characterization, and X-Ray Crystal Structures of Cucurbit[n]uril (n = 5, 7 and 8). *J. Am. Chem. Soc.*, 122, pp. 540–541.

20. Day, A., Arnold, A. P., Blanch, R. J. and Snushall, B. (2001). Controlling Factors in the Synthesis of Cucurbituril and Its Homologues. *J. Org. Chem.*, 66, pp. 8094–8100.

21. Lee, J. W., Samal, S., Selvapalam, N., Kim, H. J. and Kim, K. (2003). Cucurbituril Homologues and Derivatives: New Opportunities in Supramolecular Chemistry. *Acc. Chem. Res.*, 36, pp. 621–630.

22. Kim, H.-J., Jeon, W. S., Ko, Y. H. and Kim, K. (2002). Inclusion of Methylviologen in Cucurbit[7]uril. *Proc. Natl. Acad. Sci. U.S.A.*, 99, pp. 5007–5011.

23. Ong, W., Gomez-Kaifer, M. and Kaifer, A. E. (2002). Cucurbit[7]uril: A Very Effective Host for Viologens and Their Cation Radicals. *Org. Lett.*, 4, pp. 1791–1794.

24. Chinai, J. M., Taylor, A. B., Ryno, L. M., Hargreaves, N. D., Morris, C. A., Hart, P. J. and Urbach, A. R. (2011). Molecular Recognition of Insulin by a Synthetic Receptor. *J. Am. Chem. Soc.*, 133, pp. 8810–8813.

25. Ghale, G. and Nau, W. M. (2014). Dynamically Analyte-Responsive Macrocyclic Host–Fluorophore Systems. *Acc. Chem. Res.*, 47, pp. 2150–2159.

26. Kim, H. J., Heo, J., Jeon, W. S., Lee, E., Kim, J., Sakamoto, S. Yamaguchi, K. and Kim, K. (2001). Selective Inclusion of a Hetero–Guest Pair in a Molecular Host: Formation of Stable Charge-Transfer Complexes in Cucurbit[8]uril. *Angew. Chem. Int. Ed.*, 40, pp. 1526–1529.

27. Liu, Y., Yu, Y., Gao, J., Wang, Z. and Zhang, X. (2010). Water-Soluble Supramolecular Polymerization Driven by Multiple Host-Stabilized Charge-Transfer Interactions. *Angew. Chem. Int. Ed.*, 49, pp. 6576–6579.

28. Rauwald, U. and Scherman, O. A. (2008). Supramolecular Block Copolymers with Cucurbit[8]uril in Water. *Angew. Chem. Int. Ed.*, 47, pp. 3950–3953.

29. Jeon, Y. J., Bharadwaj, P. K., Choi, S., Lee, J. W. and Kim, K. (2002). Supramolecular Amphiphiles: Spontaneous Formation of Vesicles Triggered by Formation of a Charge-Transfer Complex in a Host. *Angew. Chem. Int. Ed.*, 41, pp. 4474–4476.

30. Liu, S., Ruspic, C., Mukhopadhyay, P., Chakrabarti, S., Zavalij, P. Y. and Isaacs, L. (2005). The Cucurbit[*n*]uril Family: Prime Components for Self-Sorting Systems. *J. Am. Chem. Soc.*, 127, pp.15959–15967.

31. Jeon, W. S., Moon, K., Park, S. H., Chun, H., Ko, Y. H., Lee, J. Y., Lee, E. S., Samal, S., Selvapalam, N., Rekharsky, M. V. Sindelar, V., Sobransingh, D., Inoue, Y., Khifer, A. E. and Kim, K. (2005). Complexation of Ferrocene Derivatives by the Cucurbit[7]uril Host: A Comparative Study of the Cucurbituril and Cyclodextrin Host Families. *J. Am. Chem. Soc.*, 127, pp. 12984–12989.

32. Rekharsky, M. V., Mori, T., Yang, C., Ko, Y. H., Selvapalam, N., Kim, H., Sobransingh, D., Kaifer, A. E., Liu, S., Isaacs, L. Isaacs, L., Chen, W., Moghaddam, S., Gilson, M. K. and Inoue, Y. (2007). A Synthetic Host-Guest System Achieves Avidin-Biotin Affinity by Overcoming Enthalpy–Entropy Compensation. *Proc. Natl. Acad. Sci.*, 104, pp. 20737–20742.

33. Cao, L., Šekutor, M., Zavalij, P. Y., Mlinarič-Majerski, K., Glaser, R. and Isaacs, L. (2014). Cucurbit[7]uril-Guest Pair with an Attomolar Dissociation Constant. *Angew. Chem. Int. Ed.*, 53, pp. 988–993.

34. Appel, E. A., Biedermann, F., Rauwald, U., Jones, S. T., Zayed, J. M. and Scherman, O. A. (2010). Supramolecular Cross-Linked Networks via Host–Guest Complexation with Cucurbit[8]uril. *J. Am. Chem. Soc.*, 132, pp. 14251–14260.

35. Park, K. M., Yang, J., Jung, H., Yeom, J., Park, J. S., Park, K., Hoffman, A. S., Hahn, S. K. and Kim, K. (2012). *In Situ* Supramolecular Assembly and Modular Modification of Hyaluronic Acid Hydrogels for 3D Cellular Engineering. *ACS Nano*, 6, pp. 2960–2968.

36. Kim, C., Agasti, S. S., Zhu, Z., Isaacs, L. and Rotello, V. M. (2010). Recognition-Mediated Activation of Therapeutic Gold Nanoparticles inside Living Cells. *Nat. Chem.*, 2, pp. 962–966.

37. Jon, S. Y., Selvapalam, N., Oh, D. H., Kang, J., Kim, S., Jeon, Y. J., Lee J. W. and Kim, K. (2003). Facile Synthesis of Cucurbit[n]uril Derivatives via Direct Functionalization: Expanding Utilization of Cucurbit[n]uril. *J. Am. Chem. Soc.*, 125, pp. 10186–10187.

38. Vinciguerra, B., Cao, L., Cannon, J. R., Zavalij, P. Y., Fenselau, C. and Isaacs, L. (2012). Synthesis and Self-Assembly Processes of Monofunctionalized Cucurbit[7]uril. *J. Am. Chem. Soc.*, 134, pp. 13133–13140.

39. Ayhan, M. M., Karoui, H., Hardy, M., Rockenbauer, A., Charles, L., Rosas, R., Udachin, K., Tordo, P., Bardelang, D. and Ouari, O. (2015). Comprehensive Synthesis of Monohydroxy-Cucurbit[n]urils (n = 5, 6, 7, 8): High Purity and High Conversions. *J. Am. Chem. Soc.*, 137, pp. 10238–10245.

40. Mock, W. L., Irra, T. A., Wepsiec, J. P. and Manimaran, T. L. (1983). Cycloaddition Induced by Cucurbituril. A Case of Pauling Principle Catalysis. *J. Org. Chem.*, 48, pp. 3619–3620.

41. Mock, W. L. and Shih, N.-Y. (1986). Structure and Selectivity in Host–Guest Complexes of Cucurbituril. *J. Org. Chem.*, 51, pp. 4440–4446.

42. Flinn, D. A., Hough, G. C., Stoddart, P. J. F. and Williams, D. D. J. (1992). Decamethylcucurbit[5]uril. *Angew. Chem. Int. Ed.*, 31, pp. 1475–1477.

43. Hoffmann, R., Knoche, W. and Fenn, C. (1994). Host–Guest Complexes of Cucurbituril with the 4-Methylbenzylammonium Ion, Alkali–Metal Cations and NH_4^+. *J. Chem. Soc. Faraday Trans.*, 90, pp. 1507–1511.

44. Whang, D., Jeon, Y., Heo, J. and Kim, K. (1996). Self-Assembly of a Polyrotaxane Containing a Cyclic "Bead" in Every Structural Unit in the Solid State: Cucurbituril Molecules Threaded on a One-Dimensional Coordination Polymer. *J. Am. Chem. Soc.*, 118, pp. 11333–11334.

45. Karcher, S., Kornmüller, A. and Jekel, M. (1999). Effects of Alkali and Alkaline–Earth Cations on the Removal of Reactive Dyes with Cucurbituril. *Acta Hydrochim. Hydrobiol.*, 27, pp. 38–42.

46. Marquez, C. and Nau, W. M. (2001). Polarizabilities Inside Molecular Containers. *Angew. Chem. Int. Ed.*, 40, pp. 4387–4390.
47. Haouaj, M., El Ho Ko, Y., Luhmer, M., Kim, K. and Bartik, K. (2001). NMR Investigation of the Complexation of Neutral Guests by Cucurbituril. *J. Chem. Soc. Perkin Trans.*, 2, pp. 2104–2107.
48. Kellersberger, K. A., Anderson, J. D., Ward, S. M., Krakowiak, K. E. and Dearden, D. V. (2001). Encapsulation of N_2, O_2, Methanol, or Acetonitrile by Decamethylcucurbit[5]uril($NH4^+$)$_2$ Complexes in the Gas Phase: Influence of the Guest On "Lid" Tightness. *J. Am. Chem. Soc.*, 123, pp. 11316–11317.
49. Day, A. I., Blanch, R. J., Arnold, A. P., Lorenzo, S., Lewis, G. R. and Dance, I. (2002). A Cucurbituril-Based Gyroscane. *Angew. Chem. Int. Ed.*, 41, pp. 275–277.
50. Zhang, H., Paulsen, E. S., Walker, K. A., Krakowiak, K. E. and Dearden, D. V. (2003). Cucurbit[6]uril Pseudorotaxanes: Distinctive Gas-Phase Dissociation and Reactivity. *J. Am. Chem. Soc.*, 125, pp. 9284–9285.
51. Miyahara, Y., Abe, K. and Inazu, T. (2002). "Molecular" Molecular Sieves: Lid-Free Decamethylcucurbit[5]uril Absorbs and Desorbs Gases Selectively. *Angew. Chem. Int. Ed.*, 41, pp. 3020–3023.
52. Miyahara, Y., Goto, K., Oka, M. and Inazu, T. (2004). Remarkably Facile Ring-Size Control in Macrocyclization: Synthesis of Hemicucurbit[6]uril and Hemicucurbit[12]uril. *Angew. Chem. Int. Ed.*, 43, pp. 5019–5022.
53. Márquez, C., Hudgins, R. R. and Nau, W. M. (2004). Mechanism of Host–Guest Complexation by Cucurbituril. *J. Am. Chem. Soc.*, 126, pp. 5806–5816.
54. Mukhopadhyay, P., Wu, A. and Isaacs, L. (2004). Social Self-Sorting in Aqueous Solution. *J. Org. Chem.*, 69, pp. 6157–6164.
55. Sasmal, S., Sinha, M. K. and Keinan, E. (2004). Facile Purification of Rare Cucurbiturils by Affinity Chromatography. *Org. Lett.*, 6, pp. 1225–1228.
56. Jeon, W. S., Kim, E., Ko, Y. H., Hwang, I., Lee, J. W., Kim, S. Y., Kim, H. J. and Kim, K. (2004). Molecular Loop Lock: A Redox-Driven Molecular Machine Based on a Host-Stabilized Charge-Transfer Complex. *Angew. Chem. Int. Ed.*, 44, pp. 87–91.
57. Isaacs, L., Park, S. K., Liu, S., Ko, Y. H., Selvapalam, N., Kim, Y., Kim, H., Zavalij, P. Y. Kim, G. H., Lee, H. S. and Kim, K. (2005). The Inverted Cucurbit[n]uril Family. *J. Am. Chem. Soc.*, 127, pp. 18000–18001.
58. Liu, S., Zavalij, P. Y. and Isaacs, L. (2005). Cucurbit[10]uril. *J. Am. Chem. Soc.*, 127, pp. 16798–16799.

59. Mohanty, J. and Nau, W. M. (2005). Ultrastable Rhodamine with Cucurbituril. *Angew. Chem. Int. Ed.*, 44, pp. 3750–3754.

60. Bush, M. E., Bouley, N. D. and Urbach, A. R. (2005). Charge-Mediated Recognition of N-Terminal Tryptophan in Aqueous Solution by a Synthetic Host. *J. Am. Chem. Soc.*, 127, pp. 14511–14517.

61. Rekharsky, M. V., Yamamura, H., Inoue, C., Kawai, M., Osaka, I., Arakawa, R., Shiba, K., Sato, A., Young, H. K., Selvapalam, N., Kim, K. and Inoue, Y. (2006). Chiral Recognition in Cucurbituril Cavities. *J. Am. Chem. Soc.*, 128, pp. 14871–14880.

62. Huang, W. H., Liu, S., Zavalij, P. Y. and Isaacs, L. (2006). Nor-Seco-Cucurbit[10]uril Exhibits Homotropic Allosterism. *J. Am. Chem. Soc.*, 128, pp. 14744–14745.

63. Heitmann, L. M., Taylor, A. B., Hart, P. J. and Urbach, A. R. (2006). Sequence-Specific Recognition and Cooperative Dimerization of N-Terminal Aromatic Peptides in Aqueous Solution by a Synthetic Host. *J. Am. Chem. Soc.*, 128, pp. 12574–12581.

64. Wang, W. and Kaifer, A. E. (2006). Electrochemical Switching and Size Selection in Cucurbit[8]uril-Mediated Dendrimer Self-Assembly. *Angew. Chem. Int. Ed.*, 45, pp. 7042–7046.

65. Sobransingh, D. and Kaifer, A. E. (2006). Electrochemically Switchable Cucurbit[7]uril-Based Pseudorotaxanes. *Org. Lett.*, 8, pp. 3247–3250.

66. Mukhopadhyay, P., Zavalij, P. Y. and Isaacs, L. (2006). High Fidelity Kinetic Self-Sorting in Multi-Component Systems Based on Guests with Multiple Binding Epitopes. *J. Am. Chem. Soc.*, 128, pp. 14093–14102.

67. Nagarajan, E. R., Oh, D. H., Selvapalam, N., Ko, Y. H., Park, K. M. and Kim, K. (2006). Cucurbituril Anchored Silica Gel. *Tetrahedron Lett.*, 47, pp. 2073–2075.

68. Huang, W. H., Zavalij, P. Y. and Isaacs, L. (2007). Chiral Recognition Inside a Chiral Cucurbituril. *Angew. Chem. Int. Ed.*, 46, pp. 7425–7427.

69. Hwang, I., Baek, K., Jung, M., Kim, Y., Park, K. M., Lee, D.-W., Selvapalam, N. and Kim, K. (2007). Noncovalent Immobilization of Proteins on a Solid Surface by Cucurbit[7]uril-Ferrocenemethylammonium Pair, a Potential Replacement of Biotin-Avidin Pair. *J. Am. Chem. Soc.*, 129, pp. 4170–4171.

70. Hwang, I., Jeon, W. S., Kim, H. J., Kim, D., Kim, H., Selvapalam, N., Fujita, N., Shinkai, S. and Kim, K. (2007). Cucurbit[7]uril: A Simple Macrocyclic, pH-Triggered Hydrogelator Exhibiting Guest-Induced Stimuli-Responsive Behavior. *Angew. Chem. Int. Ed.*, 46, pp. 210–213.

71. Kim, D., Kim, E., Kim, J., Park, C. G., Na, O. S., Lee, D.-K., Lee, K. E., Han, S. S. and Kim, K. (2007). Direct Synthesis of Polymer Nanocapsules

with a Noncovalently Tailorable Surface. *Angew. Chem. Int. Ed.*, 46, pp. 3471–3474.

72. Hennig, A., Bakirci, H. and Nau, W. M. (2007). Label-Free Continuous Enzyme Assays with Macrocycle-Fluorescent Dye Complexes. *Nat. Methods*, 4, pp. 629–632.

73. Huang, W., Zavalij, P. Y. and Isaacs, L. (2008). Cucurbit[*n*]uril Formation Proceeds by Step-Growth Cyclo-Oligomerization. *J. Am. Chem. Soc.*, 130, pp. 8446–8454.

74. Angelos, S., Yang, Y. W., Patel, K., Stoddart, J. F. and Zink, J. I. (2008). pH-Responsive Supramolecular Nanovalves Based on Cucurbit[6]uril Pseudorotaxanes. *Angew. Chem. Int. Ed.*, 47, pp. 2222–2226.

75. Lim, S., Kim, H., Selvapalam, N., Kim, K. J., Cho, S. J., Seo, G. and Kim, K. (2008). Cucurbit[6]uril: Organic Molecular Porous Material with Permanent Porosity, Exceptional Stability, and Acetylene Sorption Properties. *Angew. Chem. Int. Ed.*, 47, pp. 3352–3355.

76. An, Q., Li, G., Tao, C., Li, Y., Wu, Y. and Zhang, W. (2008). A General and Efficient Method to Form Self-Assembled Cucurbit[*n*]uril Monolayers on Gold Surfaces. *Chem. Comm.*, 44, pp. 1989–1991.

77. Megyesi, M., Biczk, L. and Jablonkai, I. (2008). Highly Sensitive Fluorescence Response to Inclusion Complex Formation of Berberine Alkaloid with Cucurbit[7]uril. *J. Phys. Chem. C*, 112, pp. 3410–3416.

78. Liu, Y., Shi, J., Chen, Y. and Ke, C. F. (2008). A Polymeric Pseudorotaxane Constructed from Cucurbituril and Aniline, and Stabilization of Its Radical Cation. *Angew. Chem. Int. Ed.*, 47, pp. 7293–7296.

79. Masson, E., Lu, X., Ling, X. and Patchell, D. L. (2009). Kinetic vs Thermodynamic Self-Sorting of Cucurbit[6]uril, Cucurbit[7]uril, and a Spermine Derivative. *Org. Lett.*, 11, pp. 2007–2010.

80. Svec, J., Necas, M. and Sindelar, V. (2010). Bambus[6]uril. *Angew. Chem. Int. Ed.*, 49, pp. 2378–2381.

81. Uzunova, V. D., Cullinane, C., Brix, K., Nau, W. M. and Day, A. I. (2010). Toxicity of Cucurbit[7]uril and Cucurbit[8]uril: An Exploratory *In Vitro* and *In Vivo* Study. *Org. Biomol. Chem.*, 8, pp. 2037–2042.

82. Hettiarachchi, G., Nguyen, D., Wu, J., Lucas, D., Ma, D., Isaacs, L. and Briken, V. (2010). Toxicology and Drug Delivery by Cucurbit[*n*]uril Type Molecular Containers. *PLoS One*, 5 e10514.

83. Florea, M. and Nau, W. M. (2011). Strong Binding of Hydrocarbons to Cucurbituril Probed by Fluorescent Dye Displacement: A Supramolecular Gas-Sensing Ensemble. *Angew. Chem. Int. Ed.*, 50, pp. 9338–9342.

84. Lee, D.-W., Park, K. M., Banerjee, M., Ha, S. H., Lee, T., Suh, K., Paul, S., Jung, H., Kim, J., Selvapalam, N., Ryu, S. H. and Kim, K. (2011).

Supramolecular Fishing for Plasma Membrane Proteins Using an Ultrastable Synthetic Host-Guest Binding Pair. *Nat. Chem.*, 3, pp. 154–159.

85. Taylor, R. W., Lee, T. C., Scherman, O. A., Esteban, R., Aizpurua, J., Huang, F. M., Baumberg, J. J. and Mahajan, S. (2011). Precise Subnanometer Plasmonic Junctions for SERS within Gold Nanoparticle Assemblies Using Cucurbit[n]uril "Glue." *ACS Nano*, 5, pp. 3878–3887.

86. Uhlenheuer, D. A., Young, J. F., Nguyen, H. D., Scheepstra, M. and Brunsveld, L. (2011). Cucurbit[8]uril Induced Heterodimerization of Methylviologen and Naphthalene Functionalized Proteins. *Chem. Commun.*, 47, pp. 6798–6800.

87. Jiang, W., Wang, Q., Linder, I., Klautzsch, F. and Schalley, C. A. (2011). Self-Sorting of Water-Soluble Cucurbituril Pseudorotaxanes. *Chem. Eur. J.*, 17, pp. 2344–2348.

88. Zhang, J., Coulston, R., Jones, S., Geng, J., Scherman, O. and Abell, C. (2012). One-Step Fabrication of Supramolecular Microcapsules from Microfluidic Droplets. *Science*, 335, pp. 690–694.

89. Biedermann, F., Uzunova, V. D., Scherman, O. A., Nau, W. M. and De Simone, A. (2012). Release of High-Energy Water as an Essential Driving Force for the High-Affinity Binding of Cucurbit[n]urils. *J. Am. Chem. Soc.*, 134, 15318–15323.

90. Shen, C., Ma, D., Meany, B., Isaacs, L. and Wang, Y. (2012). Acyclic Cucurbit[n]uril Molecular Containers Selectively Solubilize Single-Walled Carbon Nanotubes in Water. *J. Am. Chem. Soc.*, 134, pp. 7254–7257.

91. Ke, C., Smaldone, R. A., Kikuchi, T., Li, H., Davis, A. P. and Stoddart, J. F. (2013). Quantitative Emergence of Hetero[4]rotaxanes by Template-Directed Click Chemistry. *Angew. Chem. Int. Ed.*, 52, pp. 381–387.

92. An, Q., Brinkmann, J., Huskens, J., Krabbenborg, S., De Boer, J. and Jonkheijm, P. (2012). A Supramolecular System for the Electrochemically Controlled Release of Cells. *Angew. Chem. Int. Ed.*, 51, pp. 12233–12237.

93. Plumb, J. A., Venugopal, B., Oun, R., Gomez-Roman, N., Kawazoe, Y., Venkataramanan, N. S. and Wheate, N. J. (2012). Cucurbit[7]uril Encapsulated Cisplatin Overcomes Cisplatin Resistance via a Pharmacokinetic Effect. *Metallomics*, 4, pp. 561–567.

94. Baek, K., Yun, G., Kim, Y., Kim, D., Hota, R., Hwang, I., Xu, D., Ko, Y. H., Gu, G. H., Suh, J. H. Suh, J. H., Park, C. G., Sung, B. J. and Kim, K. (2013). Free-Standing, Single-Monomer-Thick Two-Dimensional Polymers through Covalent Self-Assembly in Solution. *J. Am. Chem. Soc.*, 135, pp. 6523–6528.

95. Ahn, Y., Jang, Y., Selvapalam, N., Yun, G. and Kim, K. (2013). Supramolecular Velcro for Reversible Underwater Adhesion. *Angew. Chem. Int. Ed.*, 52, pp. 3140–3144.

96. Zhang, K., Da Tian, J., Hanifi, D., Zhang, Y., Sue, A. C. H., Zhou, T. Y., Zhang, L., Zhao, X., Liu, Y. and Li, Z. T. (2013). Toward a Single-Layer Two-Dimensional Honeycomb Supramolecular Organic Framework in Water. *J. Am. Chem. Soc.*, 135, pp. 17913–17918.

97. Dang, D. T., Nguyen, H. D., Merkx, M. and Brunsveld, L. (2013). Supramolecular Control of Enzyme Activity through Cucurbit[8]uril-Mediated Dimerization. *Angew. Chem. Int. Ed.*, 52, pp. 2915–2919.

98. Cao, L., Hettiarachchi, G., Briken, V. and Isaacs, L. (2013). Cucurbit[7]uril Containers for Targeted Delivery of Oxaliplatin to Cancer Cells. *Angew. Chem. Int. Ed.*, 52, pp. 12033–12037.

99. Cheng, X. J., Liang, L. L., Chen, K., Ji, N. N., Xiao, X., Zhang, J. X., Zhang, Y. Q., Xue, S. F., Zhu, Q. J., Ni, X. L. and Tao, Z. (2013). Twisted Cucurbit[14]uril. *Angew. Chem. Int. Ed.*, 52, pp. 7252–7255.

100. Cera, L. and Schalley, C. A. (2014). Stimuli-Induced Folding Cascade of a Linear Oligomeric Guest Chain Programmed through Cucurbit[n]uril Self-Sorting (n = 6, 7, 8). *Chem. Sci.*, 5, pp. 2560–2567.

101. Joseph, R., Nkrumah, A., Clark, R. J. and Masson, E. (2014). Stabilization of Cucurbituril/Guest Assemblies via Long-Range Coulombic and CH···O Interactions. *J. Am. Chem. Soc.*, 136, pp. 6602–6607.

102. Tian, J., Zhou, T.-Y., Zhang, S.-C., Aloni, S., Altoe, M. V., Xie, S.-H., Wang, H. Zhang, D.-W., Zhao, X., Liu, Y. and Li, Z.-T. (2014). Three-Dimensional Periodic Supramolecular Organic Framework Ion Sponge in Water and Microcrystals. *Nat. Commun.*, 5, p. 5574.

103. Gong, B., Choi, B. K., Kim, J. Y., Shetty, D., Ko, Y. H., Selvapalam, N., Lee, N. K. and Kim, K. (2015). High Affinity Host–Guest FRET Pair for Single-Vesicle Content-Mixing Assay: Observation of Flickering Fusion Events. *J. Am. Chem. Soc.*, 137, pp. 8908–8911.

104. Kim, J., Baek, K., Shetty, D., Selvapalam, N., Yun, G., Kim, N. H., Ko, Y. H., Park, K. M., Hwang, I. and Kim, K. (2015). Reversible Morphological Transformation between Polymer Nanocapsules and Thin Films through Dynamic Covalent Self-Assembly. *Angew. Chem. Int. Ed.*, 54, pp. 2693–2697.

105. Tonga, G. Y., Jeong, Y., Duncan, B., Mizuhara, T., Mout, R., Das, R., Kim, S. T., Yeh, Y.-C., Yan, B., Hou, S. and Rotello, V. M. (2015). Supramolecular Regulation of Bioorthogonal Catalysis in Cells Using Nanoparticle-Embedded Transition Metal Catalysts. *Nat. Chem.*, 7, pp. 597–603.

106. Zheng, L., Sonzini, S., Ambarwati, M., Rosta, E., Scherman, O. A. and Herrmann, A. (2015). Turning Cucurbit[8]uril into a Supramolecular Nanoreactor for Asymmetric Catalysis. *Angew. Chem. Int. Ed.*, 54, pp. 13007–13011.

107. Lu, X. and Isaacs, L. (2016). Uptake of Hydrocarbons in Aqueous Solution by Encapsulation in Acyclic Cucurbit[*n*]uril-Type Molecular Containers. *Angew. Chem. Int. Ed.*, 55, pp. 8076–8080.

108. Li, Q., Qiu, S.-C., Zhang, J., Chen, K., Huang, Y., Xiao, X., Zhang, Y., Li, F., Zhang, Y.-Q., Xue, S.-F. Zhu, Q.-J., Tao, Z., Lindoy, L. F. and Wei, G. (2016). Twisted Cucurbit[*n*]urils. *Org. Lett.*, 18, pp. 4020–4023.

109. Bruns, C. J., Liu, H. and Francis, M. B. (2016). Near-Quantitative Aqueous Synthesis of Rotaxanes via Bioconjugation to Oligopeptides and Proteins. *J. Am. Chem. Soc.*, 138, pp. 15307–15310.

110. Finbloom, J. A., Han, K., Slack, C. C., Furst, A. L. and Francis, M. B. (2017). Cucurbit[6]uril-Promoted Click Chemistry for Protein Modification. *J. Am. Chem. Soc.*, 139, pp. 9691–9697.

111. Sigwalt, D., Šekutor, M., Cao, L., Zavalij, P. Y., Hostaš, J., Ajani, H., Hobza, P., Mlinarić-Majerski, K., Glaser, R. and Isaacs, L. (2017). Unraveling the Structure–Affinity Relationship between Cucurbit[*n*]urils (*n* = 7, 8) and Cationic Diamondoids. *J. Am. Chem. Soc.*, 139, pp. 3249–3258.

112. Palma, A., Artelsmair, M., Wu, G., Lu, X., Barrow, S. J., Uddin, N., Rosta, E., Masson, E. and Scherman, O. A. (2017). Cucurbit[7]uril as a Supramolecular Artificial Enzyme for Diels–Alder Reactions. *Angew. Chem. Int. Ed.*, 56, pp. 15688–15692.

113. Buschmann, H.-J. (1997). Preparation of Cucurbituril. DE 19603377 A1.

114. Marquez, C., Huang, F. and Nau, W. M. (2004). Cucurbiturils: Molecular Nanocapsules for Time-Resolved Fluorescence-Based Assays. *IEEE Trans. Nanobioscience*, 3, pp. 39–45.

115. Kim, K., Samal, S., Raju, N. K., Selvapalam, N. and Oh, D. H. (2005). Processes of Preparing Glycolurils and Cucurbiturils Using Microwaves. WO 2005103053 A1.

116. Wheate, N. J., Patel, N. and Sutcliffe, O. B. (2010). Microwave Synthesis of Cucurbit[*n*]urils. *Future Med. Chem.*, 2, pp. 231–236.

117. Yang, Q., Li, X. L., Jiang, Y., Hu, L. Y. and Yang, Y. L. (2014). Microwave Synthesis, Charaterisation and Electrochemical Property of Cucurbit[*n*] urils. *Mater. Res. Innov.*, 18, pp. 280–283.

118. Cong, H., Ni, X. L., Xiao, X., Huang, Y., Zhu, Q.-J., Xue, S.-F., Tao, Z. Lindoy, L. F. and Wei, G. (2016). Synthesis and Separation of Cucurbit[*n*] urils and Their Derivatives. *Org. Biomol. Chem.*, 14, pp. 4335–4364.

119. Jiao, D., Zhao, N. and Scherman, O. A. (2010). A "Green" Method for Isolation of Cucurbit[7]uril via a Solid State Metathesis Reaction. *Chem. Commun.*, 46, pp. 2007–2009.
120. Yi, S. and Kaifer, A. E. (2011). Determination of the Purity of Cucurbit[*n*] uril (*n* = 7, 8) Host Samples. *J. Org. Chem.*, 76, pp. 10275–10278.
121. Witt, D., Lagona, J., Damkaci, F., Fettinger, J. C. and Isaacs, L. (2000). Diastereoselective Formation of Methylene-Bridged Glycoluril Dimers. *Org. Lett.*, 2, pp. 755–758.
122. Wu, A., Mukhopadhyay, P., Chakraborty, A., Fettinger, J. C. and Isaacs, L. (2004). Molecular Clips Form Isostructural Dimeric Aggregates from Benzene to Water. *J. Am. Chem. Soc.*, 126, pp. 10035–10043.
123. Chakraborty, A., Wu, A., Witt, D., Lagona, J., Fettinger, J. C. and Isaacs, L. (2002). Diastereoselective Formation of Glycoluril Dimers: Isomerization Mechanism and Implications for Cucurbit[*n*]uril Synthesis. *J. Am. Chem. Soc.*, 124, pp. 8297–8306.
124. Liu, S., Kim, K. and Isaacs, L. (2007). Mechanism of the Conversion of Inverted CB[6] to CB[6]. *J. Org. Chem.*, 72, pp. 6840–6847.
125. Isaacs, L. (2011). The Mechanism of Cucurbituril Formation. *Isr. J. Chem.*, 51, pp. 578–591.
126. Isaacs, L. (2011). Cucurbit[*n*]urils: From Mechanism to Structure and Function. *Chem. Commun.*, 47, pp. 619–629.
127. Liu, Q., Li, Q., Cheng, X.-J., Xi, Y.-Y., Xiao, B., Xiao, X., Tang, Q., Huang, Y., Tao, Z., Xue, S.-F., Zhu, Q. J. and Zhang, J.-X. (2015). A Novel Shell-Like Supramolecular Assembly of 4,4 0-Bipyridyl Derivatives and a Twisted Cucurbit[14]uril Molecule. *Chem. Commun.*, 51, pp. 9999–10001.
128. Bardelang, D., Udachin, K. A., Leek, D. M., Margeson, J. C., Chan, G., Ratcliffe, C. I. and Ripmeester, J. A. (2011). Cucurbit[*n*]urils (*n* = 5–8): A Comprehensive Solid State Study. *Cryst. Growth Des.*, 11, pp. 5598–5614.
129. Senler, S., Li, W., Tootoonchi, M. H., Yi, S. and Kaifer, A. E. (2014). The Cucurbituril "Portal" Effect. *Supramol. Chem.*, 26, pp. 677–683.
130. Assaf, K. I. and Nau, W. M. (2014). Cucurbiturils as Fluorophilic Receptors. *Supramol. Chem.*, 26, pp. 657–669.
131. Czar, M. F. and Jockusch, R. A. (2013). Understanding Photophysical Effects of Cucurbituril Encapsulation: A Model Study with Acridine Orange in the Gas Phase. *ChemPhysChem*, 14, pp. 1138–1148.
132. Assaf, K. I. and Nau, W. M. (2015). Cucurbiturils: From Synthesis to High-Affinity Binding and Catalysis. *Chem. Soc. Rev.*, 44, pp. 394–418.

133. Nau, W. M., Florea, M. and Assaf, K. I. (2011). Deep inside Cucurbiturils: Physical Properties and Volumes of Their Inner Cavity Determine the Hydrophobic Driving Force for Host–Guest Complexation. *Isr. J. Chem.*, 51, pp. 559–577.

134. Zhao, J. Kim, H. Oh, J. Kim, S. Lee, J. W. Sakamoto, S. Yamaguchi, K. Kim, K. (2001). Cucurbit[*n*]uril Derivatives Soluble in Water and Organic Solvents. *Angew. Chem. Int. Ed.*, 40, pp. 4363–4365.

135. Wu, F., Wu, L., Xiao, X., Zhang, Y., Xue, S., Tao, Z. and Day, A. I. (2012). Locating the Cyclopentano Cousins of the Cucurbit[*n*]uril Family. *J. Org. Chem.*, 77, pp. 606–611.

136. Isobe, H., Sato, S. and Nakamura, E. (2002). Synthesis of Disubstituted Cucurbit[6]uril and Its Rotaxane Derivative. *Org. Lett.*, 4, pp. 1287–1289.

137. Wu, L. H., Ni, X. L., Wu, F., Zhang, Y. Q., Zhu, Q. J., Xue, S. F. and Tao, Z. (2009). Crystal Structures of Three Partially Cyclopentano-Substituted Cucurbit[6]urils. *J. Mol. Struct.*, 920, pp. 183–188.

138. Kim, K., Lee, J. W., Oh, D. H. and Ju, J. (2005), Distributed Cucurbiturils and Preparing Method Therof. Patents WO 2005087777 A1.

139. Lucas, D., Minami, T., Iannuzzi, G., Cao, L., Wittenberg, J. B., Anzenbacher, P. and Isaacs, L. (2011). Templated Synthesis of Glycoluril Hexamer and Monofunctionalized Cucurbit[6]uril Derivatives. *J. Am. Chem. Soc.*, 133, pp. 17966–17976.

140. Cao, L. and Isaacs, L. (2012). Daisy Chain Assembly Formed from a Cucurbit[6]uril Derivative. *Org. Lett.*, 14, pp. 3072–3075.

141. Niele, F. G. and Nolte, R. J. M. (1988). Palladium(II) Cage Compounds Based on Diphenylglycoluril. *J. Am. Chem. Soc.*, 110, pp. 172–177.

142. Day, A. I., Arnold, A. P. and Blanch, R. J. (2003). A Method for Synthesizing Partially Substituted Cucurbit[*n*]uril. *Molecules*, 8, pp. 74–84.

143. Zhao, Y., Wue, S., Zhu, Q., Tao, Z., Zhang, J., Wei Z., Long, L., Hu, M., Xiao, H. and Day. A. (2004). Synthesis of a Symmetrical Tetrasubstituted Cucurbit[6]uril and Its Host–Guest Inclusion Complex with 2,2′-Bipyridine. *Chinese Sci. Bull.*, 49, pp. 1111–1116.

144. Robinson, E. L., Zavalij, P. Y. and Isaacs, L. (2015). Synthesis of a Disulfonated Derivative of Cucurbit[7]uril and Investigations of Its Ability to Solubilise Insoluble Drugs. *Supramol. Chem.*, 27, pp. 288–289.

145. Zhao, N., Lloyd, G. O. and Scherman, O. A. (2012). Monofunctionalised Cucurbit[6]uril Synthesis Using Imidazolium Host–Guest Complexation. *Chem. Commun.*, 48, pp. 3070–3072.

146. Ayhan, M. M., Karoui, H., Hardy, M., Rockenbauer, A., Charles, L., Rosas, R., Udachin, K., Tordo, P., Bardelang, D. and Ouari, O. (2016). Correction to

Comprehensive Synthesis of Monohydroxy–Cucurbit[*n*]urils (*n* = 5, 6, 7, 8): High Purity and High Conversions. *J. Am. Chem. Soc.*, 138, p. 2060.

147. Burnett, C. A., Witt, D., Fettinger, J. C. and Isaacs, L. (2003). Acyclic Congener of Cucurbituril: Synthesis and Recognition Properties. *J. Org. Chem.*, 68, pp. 6184–6191.

148. Stand, M., Hodan, M. and Slndelar, V. (2009). Glycoluril Trimers: Selective Synthesis and Supramolecular Properties. *Org. Lett.*, 11, pp. 4184–4187.

149. Ma, D., Zavalij, P. Y. and Isaacs, L. (2010). Acyclic Cucurbit[*n*]uril Congeners Are High Affinity Hosts. *J. Org. Chem.*, 75, pp. 4786–4795.

150. Ma, D., Hettiarachchi, G., Nguyen, D., Zhang, B., Wittenberg, J. B., Zavalij, P. Y., Briken, V. and Isaacs, L. (2012). Acyclic Cucurbit[*n*]uril Molecular Containers Enhance the Solubility and Bioactivity of Poorly Soluble Pharmaceuticals. *Nat. Chem.*, 4, pp. 503–510.

151. Huang, W. H., Zavalij, P. Y. and Isaacs, L. (2008). Folding of Long-Chain Alkanediammonium Ions Promoted by a Cucurbituril Derivative. *Org. Lett.*, 10, pp. 2577–2580.

152. Aav, R., Shmatova, E., Reile, I., Borissova, M., Topić, F. and Rissanen, K. (2013). New Chiral Cyclohexylhemicucurbit[6]uril. *Org. Lett.*, *15*, pp. 3786–3789.

153. Prigorchenko, E., Oeren, M., Kaabel, S., Fomitsenko, M., Reile, I., Jarving, I., Tamm, T., Topic, F., Rissanen, K. and Aav, R. (2015). Template-Controlled Synthesis of Chiral Cyclohexylhemicucurbit[8]uril. *Chem. Commun.*, 51, pp. 10921–10924.

154. Havel, V., Svec, J., Wimmerova, M., Dusek, M., Pojarova, M. and Sindelar, V. (2011). Bambus[*n*]urils: A New Family of Macrocyclic Anion Receptors. *Org. Lett.*, 13, pp. 4000–4003.

155. Yawer, M. A., Havel, V. and Sindelar, V. (2015). A Bambusuril Macrocycle That Binds Anions in Water with High Affinity and Selectivity. *Angew. Chem. Int. Ed.*, 54, pp. 276–279.

156. Lisbjerg, M., Jenssen, B. M., Rasmussen, B., Nielsen, B. E., Madsen, A. and Pittelkow, M. (2014). Discovery of a Cyclic 6 + 6 Hexamer of D-Biotin and Formaldehyde. *Chem. Sci.*, 5, pp. 2647–2650.

157. Lagona, J., Fettinger, J. C. and Isaacs, L. (2003). Cucurbit[*n*]uril Analogues. *Org. Lett.*, 5, pp. 3745–3747.

158. Wagner, B. D. and MacRae, A. I. (1999). The Lattice Inclusion Compound of 1,8-ANS and Cucurbituril: A Unique Fluorescent Solid. *J. Phys. Chem. B*, 103, pp. 10114–10119.

159. Lagona, J., Fettinger, J. C. and Isaacs, L. (2005). Cucurbit[*n*]uril Analogues: Synthetic and Mechanistic Studies. *J. Org. Chem.*, 70, pp. 10381–10392.

160. Ustrnul, L., Kulhanek, P., Lizal, T. and Sindelar, V. (2015). Pressocucurbit[5] uril. *Org. Lett.*, 17, pp. 1022–1025.

161. Wu, Y., Xu, L., Shen, Y., Wang, Y., Zou, L., Wang, Q., Jiang, X., Liu, J. and Tian, H. (2017). The Smallest Cucurbituril Analogue with High Affinity for Ag^+. *Chem. Commun.*, 53, pp. 4070–4072.

162. Gilberg, L., Khan, M. S. A., Enderesova, M. and Sindelar, V. (2014). Cucurbiturils Substituted on the Methylene Bridge. *Org. Lett.*, 16, pp. 2446–2249.

163. Ma, D., Gargulakova, Z., Zavalij, P. Y., Sindelar, V. and Isaacs, L. (2010). Reasons Why Aldehydes Do Not Generally Participate in Cucurbit[*n*]uril Forming Reactions. *J. Org. Chem.*, 75, pp. 2934–2941.

164. Barooah, N., Mohanty, J., Pal, H. and Bhasikuttan, A. C. (2014). Cucurbituril-Induced Supramolecular pK_a Shift in Fluorescent Dyes and Its Prospective Applications. *Proc. Natl. Acad. Sci. India Sect. A — Phys. Sci.*, 84, pp. 1–17.

165. Koner, A. L. and Nau, W. M. (2007). Cucurbituril Encapsulation of Fluorescent Dyes. *Supramol. Chem.*, 19, pp. 55–66.

166. Saleh, N., Koner, A. L. and Nau, W. M. (2008). Activation and Stabilization of Drugs by Supramolecular pK_a Shifts: Drug-Delivery Applications Tailored for Cucurbiturils. *Angew. Chem. Int. Ed.*, 47, pp. 5398–5401.

167. Praetorius, A., Bailey, D. M., Schwarzlose, T. and Nau, W. M. (2008). Design of a Fluorescent Dye for Indicator Displacement from Cucurbiturils: A Macrocycle-Responsive Fluorescent Switch Operating through a pK_a Shift. *Org. Lett.*, 10, pp. 4089–4092.

168. Mohanty, J., Bhasikuttan, A. C., Nau, W. M. and Pal, H. (2006). Host-Guest Complexation of Neutral Red with Macrocyclic Host Molecules: Contrasting pK_a Shifts and Binding Affinities for Cucurbit[7]uril and β-Cyclodextrin. *J. Phys. Chem. B*, 110, pp. 5132–5138.

169. Buschmann, H.-J., Cleve, E., Jansen, K., Wego, A. and Schollmeyer, E. (2001). Complex Formation between Cucurbit[*n*]urils and Alkali, Alkaline Earth and Ammonium Ions in Aqueous Solution. *J. Incl. Phenom. Macrocycl. Chem.*, 40, pp. 117–120.

170. Buschmann, H. J., Cleve, E., Jansen, K. and Schollmeyer, E. (2001). Determination of Complex Stabilities with Nearly Insoluble Host Molecules: Cucurbit[5]uril, Decamethylcucurbit[5]uril and Cucurbit[6]uril as Ligands for the Complexation of Some Multicharged Cations in Aqueous Solution. *Anal. Chim. Acta*, 437, pp. 157–163.

171. Meschke, C., Buschmann, H.-J. and Schollmeyer, E. (1997). Complexes of Cucurbituril with Alkyl Mono- and Diammonium Ions in Aqueous Formic Acid Studied by Calorimetric Titrations. *Thermochim. Acta*, 297, pp. 43–48.

172. Buschmann, H.-J., Jansen, K. and Schollmeyer, E. (2000). Cucurbituril as Host Molecule for the Complexation of Aliphatic Alcohols, Acids and Nitriles in Aqueous Solution. *Thermochim. Acta*, 346, pp. 33–36.

173. Buschmann, H.-J., Jansen, K. and Schollmeyer, E. (2000). Cucurbituril and α-and β-Cyclodextrins as Ligands for the Complexation of Nonionic Surfactants and Polyethyleneglycols in Aqueous Solutions. *J. Incl. Phenom. Macrocycl. Chem.*, 37, pp. 231–236.

174. Buschmann, H. J., Mutihac, L., Mutihac, R. C. and Schollmeyer, E. (2005). Complexation Behavior of Cucurbit[6]uril with Short Polypeptides. *Thermochim. Acta*, 430, pp. 79–82.

175. Rekharsky, M. V., Ko, Y. H., Selvapalam, N., Kim, K. and Inoue, Y. (2007). Complexation Thermodynamics of Cucurbit[6]uril with Aliphatic Alcohols, Amines, and Diamines. *Supramol. Chem.*, 19, pp. 39–46.

176. Kim, Y., Kim, H., Ko, Y. H., Selvapalam, N., Rekharsky, M. V., Inoue, Y. and Kim, K. (2009). Complexation of Aliphatic Ammonium Ions with a Water-Soluble Cucurbit[6]uril Derivative in Pure Water: Isothermal Calorimetric, NMR, and X-Ray Crystallographic Study. *Chem. Eur. J.*, 15, pp. 6143–6151.

177. Huang, W. H., Zavalij, P. Y. and Isaacs, L. (2008). Cucurbit[6]uril *p*-Xylylenediammonium Diiodide Deca-Hydrate Inclusion Complex. *Acta Cryst. E*, 64, pp. 1321–1322.

178. Liu, L., Zhao, N. and Scherman, O. A. (2008). Ionic Liquids as Novel Guests for Cucurbit[6]uril in Neutral Water. *Chem. Commun.*, 44, pp. 1070–1072.

179. Danylyuk, O., Fedin, V. P. and Sashuk, V. (2013). Kinetic Trapping of the Host–Guest Association Intermediate and Its Transformation into a Thermodynamic Inclusion Complex. *Chem. Commun.*, 49, pp. 1859–1861.

180. Danylyuk, O., Fedin, V. P. and Sashuk, V. (2013). Host–Guest Complexes of Cucurbit[6]uril with Isoprenaline: The Effect of the Metal Ion on the Crystallization Pathway and Supramolecular Architecture. *Cryst. Eng. Comm.*, 15, p. 7414.

181. Fusaro, L., Locci, E., Lai, A. and Luhmer, M. (2008). NMR Study of the Reversible Trapping of SF_6 by Cucurbit[6]uril in Aqueous Solution. *J. Phys. Chem. B.*, 112, pp. 15014–15020.

182. El-Sheshtawy, H. S., Bassil, B. S., Assaf, K. I., Kortz, U. and Nau, W. M. (2012). Halogen Bonding inside a Molecular Container. *J. Am. Chem. Soc.*, 134, pp. 19935–19941.

183. Whang, D., Heo, J., Park, J. H. and Kim, K. (1998). A Molecular Bowl with Metal Ion as Bottom: Reversible Inclusion of Organic Molecules in Cesium Ion Complexed Cucurbituril. *Angew. Chem. Int. Ed.*, 37, pp. 78–80.

184. Ong, W. and Kaifer, A. E. (2003). Unusual Electrochemical Properties of the Inclusion Complexes of Ferrocenium and Cobaltocenium with Cucurbit[7]uril. *Organometallics*, 22, pp. 4181–4183.

185. Wang, R. and Macartney, D. H. (2008). Cucurbit[7]uril Stabilization of a Diarylmethane Carbocation in Aqueous Solution. *Tetrahedron Lett.*, 49, pp. 311–314.

186. Lee, J. W., Lee, H. H. L., Ko, Y. H., Kim, K. and Kim, H. I. (2015). Deciphering the Specific High-Affinity Binding of Cucurbit[7]uril to Amino Acids in Water. *J. Phys. Chem. B*, 119, pp. 4628–4636.

187. Lee, J. W., Shin, M. H., Mobley, W., Urbach, A. R. and Kim, H. I. (2015). Supramolecular Enhancement of Protein Analysis via the Recognition of Phenylalanine with Cucurbit[7]uril. *J. Am. Chem. Soc.*, 137, pp. 15322–15329.

188. Logsdon, L. and Urbach, A. R. (2013). Sequence-Specific Inhibition of a Nonspecific Protease. *J. Am. Chem. Soc.*, 7, pp. 11414–11416.

189. Jang, Y., Natarajan, R., Ko, Y. H. and Kim, K. (2014). Cucurbit[7]uril: A High-Affinity Host for Encapsulation of Amino Saccharides and Supramolecular Stabilization of Their α-Anomers in Water. *Angew. Chem. Int. Ed.*, 53, pp. 1003–1007.

190. Kim, S.-Y., Jung, I.-S., Lee, E., Kim, J., Sakamoto, S., Yamaguchi, K. and Kim, K. (2001). Macrocycles within Macrocycles: Cyclen, Cyclam, and Their Transition Metal Complexes Encapsulated in Cucurbit[8]uril. *Angew. Chem. Int. Ed.*, 40, pp. 2119–2121.

191. Ko, Y. H., Kim, H., Kim, Y. and Kim, K. (2008). U-Shaped Conformation of Alkyl Chains Bound to a Synthetic Host. *Angew. Chem. Int. Ed.*, 47, pp. 4106–4109.

192. Ko, Y. H., Kim, Y., Kim, H. and Kim, K. (2011). U-Shaped Conformation of Alkyl Chains Bound to a Synthetic Receptor Cucurbit[8]uril. *Chem. Asian J.*, 6, pp. 652–657.

193. Baek, K., Kim, Y., Kim, H., Yoon, M., Hwang, I., Ko, Y. H. and Kim, K. (2010). Unconventional U-Shaped Conformation of a Bolaamphiphile Embedded in a Synthetic Host. *Chem. Commun.*, 46, pp. 4091–4093.

194. Mileo, E., Mezzina, E., Grepioni, F., Pedulli, G. F. and Lucarini, M. (2009). Preparation and Characterisation of a New Inclusion Compound of Cucurbit[8]uril with a Nitroxide Radical. *Chem. Eur. J.*, 15, pp. 7859–7862.

195. Biedermann, F. and Scherman, O. A. (2012). Model System for Studying Charge-Transfer Interactions. *J. Phys. Chem. B*, 116, pp. 2842–2849.

196. Liu, J. X., Lin, R. L., Long, L. S., Huang, R. and Bin Zheng, L. S. (2008). A Novel Inclusion Complex Form between Q[10] Host and Q[5] Guest Stabilized by Potassium Ion Coordination. *Inorg. Chem. Commun.*, 11, pp. 1085–1087.

197. Liu, S., Shukla, A. D., Gadde, S., Wagner, B. D., Kaifer, A. E. and Isaacs, L. (2008). Ternary Complexes Comprising Cucurbit[10]uril, Porphyrins, and Guests. *Angew. Chem. Int. Ed.*, 47, pp. 2657–2660.

198. Gong, W., Yang, X., Zavalij, P. Y., Isaacs, L., Zhao, Z. and Liu, S. (2016). From Packed Sandwich to Russian Doll: Assembly by Charge-Transfer Interactions in Cucurbit[10]uril. *Chem. Eur. J.*, 22 (49), pp. 17612–17618.

199. Liu, S., Zavalij, P. Y., Lam, Y. F. and Isaacs, L. (2007). Refolding Foldamers: Triazene-Arylene Oligomers That Change Shape with Chemical Stimuli. *J. Am. Chem. Soc.*, 129, pp. 11232–11241.

200. Pisani, M. J., Zhao, Y., Wallace, L., Woodward, C. E., Keene, F. R., Day, A. I. and Collins, J. G. (2010). Cucurbit[10]uril Binding of Dinuclear Platinum(II) and Ruthenium(II) Complexes: Association/Dissociation Rates from Seconds to Hours. *Dalt. Trans.*, 39, pp. 2078–2086.

201. Alrawashdeh, L. R. Cronin, M. P. Woodward, C. E. Day, A. I. and Wallace, L. (2016). Iridium Cyclometalated Complexes in Host–Guest Chemistry: A Strategy for Maximizing Quantum Yield in Aqueous Media. *Inorg. Chem.*, 55, pp. 6759–6769.

202. Chodera, J. D. and Mobley, D. L. (2013). Entropy-Enthalpy Compensation: Role and Ramifications in Biomolecular Ligand Recognition and Design. *Annu. Rev. Biophys.*, 42, pp. 121–142.

203. Moghaddam, S., Yang, C., Rekharsky, M., Ko, Y. H., Kim, K., Inoue, Y. and Gilson, M. K. (2011). New Ultrahigh Affinity Host–Guest Complexes of Cucurbit[7]uril with Bicyclo[2.2.2]octane and Adamantane Guests: Thermodynamic Analysis and Evaluation of M2 Affinity Calculations. *J. Am. Chem. Soc.*, 133, pp. 3570–3581.

204. Ling, X., Saretz, S., Xiao, L., Francescon, J. and Masson, E. (2016). Water vs Cucurbituril Rim: A Fierce Competition for Guest Solvation. *Chem. Sci.*, 7, pp. 3569–3573.

205. Rekharsky, M. V. and Inoue, Y. (1998). Complexation Thermodynamics of Cyclodextrins. *Chem. Rev.*, 2665, pp. 4–6.

206. Biedermann, F., Nau, W. M. and Schneider, H. J. (2014). The Hydrophobic Effect Revisited — Studies with Supramolecular Complexes Imply High-Energy Water as a Noncovalent Driving Force. *Angew. Chem. Int. Ed.*, 53, pp. 11158–11171.

207. Biedermann, F., Vendruscolo, M., Scherman, O. A., De Simone, A. and Nau, W. M. (2013). Cucurbit[8]uril and Blue-Box: High-Energy Water Release Overwhelms Electrostatic Interactions. *J. Am. Chem. Soc.*, 135, pp. 14879–14888.

208. Ramalingam, V. and Urbach, A. R. (2011). Cucurbit[8]uril Rotaxanes. *Org. Lett.*, 13, pp. 4898–4901.

209. Mock, W. L. and Shih, N.-Y. (1989). Dynamics of Molecular Recognition Involving Cucurbituril. *J .Am. Chem. Soc.*, 111, pp. 2697–2699.

210. Marquez, C. and Nau, W. M. (2001). Two Mechanisms of Slow Host–Guest Complexation between Cucurbit[6]uril and Supramolecular Kinetics. *Angew. Chem. Int. Ed.*, 40, pp. 3155–3160.

211. Tang, H., Fuentealba, D., Ko, Y. H., Selvapalam, N., Kim, K. and Bohne, C. (2011). Guest Binding Dynamics with Cucurbit[7]uril in the Presence of Cations. *J. Am. Chem. Soc.*, 133, pp. 20623–20633.

212. Kalmár, J., Ellis, S. B., Ashby, M. T. and Halterman, R. L. (2012). Kinetics of Formation of the Host–Guest Complex of a Viologen with Cucurbit[7]uril. *Org. Lett.*, 14, pp. 3248–3251.

213. Miskolczy, Z. and Biczok, L. (2014). Kinetics and Thermodynamics of Berberine Inclusion in Cucurbit[7]uril. *J. Phys. Chem. B*, 118, pp. 2499–2505.

214. Lee, S. J. C., Lee, J. W., Lee, H. H., Seo, J., Noh, D. H., Ko, Y. H., Kim, K. and Kim, H. I. (2013). Host–Guest Chemistry from Solution to the Gas Phase: An Essential Role of Direct Interaction with Water for High-Affinity Binding of Cucurbit[n]urils. *J. Phys. Chem. B*, 117, pp. 8855–8864.

215. Yang, F. and Dearden, D. V. (2011). Gas Phase Cucurbit[n]uril Chemistry. *Isr. J. Chem.*, 51, pp. 551–558.

216. Dearden, D. V., Ferrell, T. A., Asplund, M. C., Zilch, L. W., Julian, R. R. and Jarrold, M. F. (2009). One Ring to Bind Them All: Shape-Selective Complexation of Phenylenediamine Isomers with Cucurbit[6]uril in the Gas Phase. *J. Phys. Chem. A*, 113, pp. 989–997.

217. Zhang, H., Ferrell, T. A., Asplund, M. C. and Dearden, D. V. (2007). Molecular Beads on a Charged Molecular String: α,ω-Alkyldiammonium Complexes of Cucurbit[6]uril in the Gas Phase. *Int. J. Mass Spectrom.*, 265, pp. 187–196.

218. Zhang, H., Grabenauer, M., Bowers, M. T. and Dearden, D. V. (2009). Supramolecular Modification of Ion Chemistry: Modulation of Peptide Charge State and Dissociation Behavior through Complexation with Cucurbit[n]uril ($n = 5, 6$) or α-Cyclodextrin. *J. Phys. Chem. A*, 113, pp. 1508–1517.

219. Heo, S. W., Choi, T. S., Park, K. M., Ko, Y. H., Kim, S. B., Kim, K. and Kim, H. I. (2011). Host–Guest Chemistry in the Gas Phase: Selected Fragmentations of CB[6]-Peptide Complexes at Lysine Residues and Its Utility to Probe the Structures of Small Proteins. *Anal. Chem.*, 83, pp. 7916–7923.

220. Lee, H. H. L., Lee, J. W., Jang, Y., Ko, Y. H., Kim, K. and Kim, H. I. (2016). Manifesting Subtle Differences of Neutral Hydrophilic Guest Isomers in a Molecular Container by Phase Transfer. *Angew. Chem. Int. Ed.*, 55 pp. 8249–8253.

221. Lee, T.-C., Kalenius, E., Lazar, A. I., Assaf, K. I., Kuhnert, N., Grün, C. H., Jänis, J., Scherman, O. A. and Nau, W. M. (2013). Chemistry Inside Molecular Containers in the Gas Phase. *Nat. Chem.*, 5, pp. 376–382.

222. Mecozzi, S. and Rebek, J. (1998). The 55% Solution: A Formula for Molecular Recognition in the Liquid State. *Chem. Eur. J.*, 4, pp. 1016–1022.

223. Buschmann, H. J., Cleve, E. and Schollmeyer, E. (2005). Hemicucurbit[6] uril, a Selective Ligand for the Complexation of Anions in Aqueous Solution. *Inorg. Chem. Commun.*, 8, pp. 125–127.

224. Buschmann, H. J., Zielesny, A. and Schollmeyer, E. (2006). Hemicucurbit[6] uril a Macrocyclic Ligand with Unusual Complexing Properties. *J. Incl. Phenom.*, 54, pp. 181–185.

225. Svec, J., Dusek, M., Fejfarova, K., Stacko, P., Klan, P., Kaifer, A. E., Li, W., Hudeckova, E. and Sindelar, V. (2011). Anion-Free Bambus[6]uril and Its Supramolecular Properties. *Chem. Eur. J.*, 17, pp. 5605–5612.

226. Solel, E., Singh, M., Reany, O. and Keinan, E. (2016). Enhanced Anion Binding by Heteroatom Replacement in Bambusurils. *Phys. Chem. Chem. Phys.*, 18, pp. 13180–13185.

227. Kolb, H. C., Finn, M. G. and Sharpless, K. B. (2001). Click Chemistry: Diverse Chemical Function from a Few Good Reactions. *Angew. Chem. Int. Ed.*, 40, pp. 2004–2021.

228. Carlqvist, P. and Maseras, F. (2007). A Theoretical Analysis of a Classic Example of Supramolecular Catalysis. *Chem. Commun.*, 43, pp. 748–750.

229. Tuncel, D., Unal, O. and Artar, M. (2011). Supramolecular Assemblies Constructed by Cucurbituril-Catalyzed Click Reaction. *Isr. J. Chem.*, 51, pp. 525–532.

230. Jon, S. Y., Ko, Y. H., Park, S. H., Kim, H.-J. and Kim, K. (2001). A Facile, Stereoselective [2 + 2] Photoreaction Mediated by Cucurbit[8]uril. *Chem. Commun.*, 37, pp. 1938–1939.

231. Wang, R., Yuan, L. and Macartney, D. H. (2006). Cucurbit[7]uril Mediates the Stereoselective [4 + 4] Photodimerization of 2-Aminopyridine Hydrochloride in Aqueous Solution. *J. Org. Chem.*, 71, pp. 1237–1239.

232. Pattabiraman, M., Natarajan, A., Kaliappan, R., Mague, T. and Ramamurthy, V. (2005). Template Directed Photodimerization of *trans*-1,2-Bis(*n*-Pyridyl)-Ethylenes and Stilbazoles in Water. *Chem. Commun.*, 41, pp. 4542–4544.

233. Pattabiraman, M., Natarajan, A., Kaanumalle, L. S. and Ramamurthy, V. (2005). Templating Photodimerization of *trans*-Cinnamic Acids with Cucurbit[8]uril and γ-Cyclodextrin. *Org. Lett.*, 7, pp. 529–532.

234. Maddipatla, M. V. S. N., Kaanumalle, L. S., Natarajan, A., Pattabiraman, M. and Ramamurthy, V. (2007). Preorientation of Olefins toward a Single Photodimer: Cucurbituril-Mediated Photodimerization of Protonated Azastilbenes in Water. *Langmuir*, 23, pp. 7545–7554.

235. Pattabiraman, M., Kaanumalle, L. S., Natarajan, A. and Ramamurthy, V. (2006). Regioselective Photodimerization of Cinnamic Acids in Water: Templation with Cucurbiturils. *Langmuir*, 22, pp. 7605–7609.

236. Pemberton, B. C., Barooah, N., Srivatsava, D. K. and Sivaguru, J. (2010). Supramolecular Photocatalysis by Confinement-Photodimerization of Coumarins within Cucurbit[8]urils. *Chem. Commun.*, 46, pp. 225–227.

237. Barooah, N., Pemberton, B. C. and Sivaguru, J. (2008). Manipulating Photochemical Reactivity of Coumarins within Cucurbituril Nanocavities. *Org. Lett.*, 10, pp. 3339–3342.

238. Yang, C., Mori, T., Origane, Y., Young, H. K., Selvapalam, N., Kim, O. and Inoue, Y. (2008). Highly Stereoselective Photocyclodimerization of α-Cyclodextrin-Appended Anthracene Mediated by γ-Cyclodextrin and Cucurbit[8]uril: A Dramatic Steric Effect Operating Outside the Binding Site. *J. Am. Chem. Soc.*, 130, pp. 8574–8575.

239. Ramamurthy, V. and Sivaguru, J. (2016). Supramolecular Photochemistry as a Potential Synthetic Tool: Photocycloaddition. *Chem. Rev.*, 116, pp. 9914–9993.

240. Basilio, N. Garcia-Rio, L. Moreira, J. A. and Pessego, M. (2010). Supramolecular Catalysis by Cucurbit[7]uril and Cyclodextrins: Similarity and Differences. *J. Org. Chem.*, 75, pp. 848–855.

241. Klock, C. Dsouza, R. N. and Nau, W. M. (2009). Cucurbituril-Mediated Supramolecular Acid Catalysis. *Org. Lett.*, 11, pp. 2595–2598.

242. De Lima, S. M., Gomez, J. A., Barros, V. P., Vertuan, G. D. S., Assis, M. D. D., Graeff, C. F. D. O. and Demets, G. J. F. (2010). A New Oxovanadium(IV)-Cucurbit[6]uril Complex: Properties and Potential for

Confined Heterogeneous Catalytic Oxidation Reactions. *Polyhedron*, 29, pp. 3008–3013.

243. Lu, X. and Masson, E. (2010). Silver-Promoted Desilylation Catalyzed by Ortho- and Allosteric Cucurbiturils. *Org. Lett.*, 12, pp. 2310–2313.

244. Koner, A. L., Márquez, C., Dickman, M. H. and Nau, W. M. (2011). Transition-Metal-Promoted Chemoselective Photoreactions at the Cucurbituril Rim. *Angew. Chem. Int. Ed.*, 50, pp. 545–548.

245. Jeon, W. S., Kim, H.-J., Lee, C. and Kim, K. (2002). Control of the Stoichiometry in Host–Guest Complexation by Redox Chemistry of Guests: Inclusion of Methylviologen in Cucurbit[8]uril. *Chem. Commun.*, 38, pp. 1828–1829.

246. Ziganshina, A. Y., Ko, Y. H., Jeon, W. S. and Kim, K. (2004). Stable π-Dimer of a Tetrathiafulvalene Cation Radical Encapsulated in the Cavity of Cucurbit[8]uril. *Chem. Commun.*, 40, pp. 806–807.

247. Hwang, I., Ziganshina, A. Y., Ko, Y. H., Yun, G. and Kim, K. (2009). A New Three-Way Supramolecular Switch Based on Redox-Controlled Interconversion of Hetero- and Homo-Guest-Pair Inclusion Inside a Host Molecule. *Chem. Commun.*, 45, pp. 416–418.

248. Eelkema, R., Maeda, K., Odell, B. and Anderson, H. L. (2007). Radical Cation Stabilization in a Cucurbituril Oligoaniline Rotaxane. *J. Am. Chem. Soc.*, 129, pp. 12384–12385.

249. Choi, S., Park, S. H., Ziganshina, A. Y., Ko, Y. H., Lee, J. W. and Kim, K. (2003). A Stable *cis*-Stilbene Derivative Encapsulated in Cucurbit[7]uril. *Chem. Commun.*, 39, pp. 2176–2177.

250. Berbeci, L. S., Wang, W. and Kaifer, A. E. (2008). Drastically Decreased Reactivity of Thiols and Disulfides Complexed by Cucurbit[6]uril. *Org. Lett*, 10, pp. 3721–3724.

251. Iali, W., Petrović, P., Pfeffer, M., Grimme, S. and Djukic, J.-P. (2012). The Inhibition of Iridium-Promoted Water Oxidation Catalysis (WOC) by Cucurbit[n]urils. *Dalt. Trans.*, 41, pp. 12233–12243.

252. Rawashdeh, A. M. M., Thangavel, A., Sotiriou-leventis, C. and Leventis, N. (2008). Control of the Ketone to Gem-Diol Equilibrium by Host–Guest Interactions. *Org. Lett.*, 10, pp. 1131–1134.

253. Macartney, D. H. (2011). Encapsulation of Drug Molecules by Cucurbiturils: Effects on Their Chemical Properties in Aqueous Solution. *Isr. J. Chem.*, 51, pp. 600–615.

254. Dsouza, R. N., Pischel, U. and Nau, W. M. (2011). Fluorescent Dyes and Their Supramolecular Host/Guest Complexes with Macrocycles in Aqueous Solution. *Chem. Rev.*, 111, pp. 7941–7980.

255. Strickler, S. J. and Berg, R. A. (1962). Relationship between Absorption Intensity and Fluorescence Lifetime of Molecules. *J. Chem. Phys.*, 37, p. 814.

256. Mohanty, J., Pal, H., Ray, A. K., Kumar, S. and Nau, W. M. (2007). Supramolecular Dye Laser with Cucurbit[7]uril in Water. *ChemPhysChem*, 8, pp. 54–56.

257. Gadde, S., Batchelor, E. K. and Kaifer, A. E. (2009). Controlling the Formation of Cyanine Dye H- and J-Aggregates with Cucurbituril Hosts in the Presence of Anionic Polyelectrolytes. *Chem. Eur. J.*, 15, pp. 6025–6031.

258. Bhasikuttan, A. C., Mohanty, J., Nau, W. M. and Pal, H. (2007). Efficient Fluorescence Enhancement and Cooperative Binding of an Organic Dye in a Supra-Biomolecular Host–Protein Assembly. *Angew. Chem. Int. Ed.*, 46, pp. 4120–4122.

259. You, L., Zha, D. and Anslyn, E. V. (2015). Recent Advances in Supramolecular Analytical Chemistry Using Optical Sensing. *Chem. Rev.*, 115, pp. 7840–7892.

260. Vazquez, J., Remon, P., Dsouza, R. N., Lazar, A. I., Arteaga, J. F., Nau, W. M. and Pischel, U. (2014). A Simple Assay for Quality Binders to Cucurbiturils. *Chem. Eur. J.*, 20, pp. 9897–9901.

261. Baumes, L. A., Sogo, M. B., Montes-Navajas, P., Corma, A. and Garcia, H. (2009). First Colorimetric Sensor Array for the Identification of Quaternary Ammonium Salts. *Tetrahedron Lett.*, 50, pp. 7001–7004.

262. Baumes, L. A., Sogo, M. B., Montes-Navajas, P., Corma, A. and Garcia, H. (2010). A Colorimetric Sensor Array for the Detection of the Date-Rape Drug γ-Hydroxybutyric Acid (GHB): A Supramolecular Approach. *Chem. Eur. J.*, 16, pp. 4489–4495.

263. Minami, T., Esipenko, N. A., Zhang, B., Kozelkova, M. E., Isaacs, L., Nishiyabu, R., Kubo, Y. and Anzenbacher, P. (2012). Supramolecular Sensor for Cancer-Associated Nitrosamines. *J. Am. Chem. Soc.*, 134, pp. 20021–20024.

264. Shcherbakova, E. G., Zhang, B., Gozem, S., Minami, T., Zavalij, P. Y., Pushina, M. Isaacs, L. and Anzenbacher, P. (2017). Supramolecular Sensors for Opiates and Their Metabolites. *J. Am. Chem. Soc.*, 139, pp. 14954–14960.

265. Nau, W. M., Ghale, G., Hennig, A., Bakirci, H. and Bailey, D. M. (2009). Substrate-Selective Supramolecular Tandem Assays: Monitoring Enzyme Inhibition of Arginase and Diamine Oxidase by Fluorescent Dye Displacement from Calixarene and Cucurbituril Macrocycles. *J. Am. Chem. Soc.*, 131, pp. 11558–11570.

266. Bailey, D. M., Hennig, A., Uzunova, V. D. and Nau, W. M. (2008). Supramolecular Tandem Enzyme Assays for Multiparameter Sensor Arrays and Enantiomeric Excess Determination of Amino Acids. *Chem. Eur. J.*, 14, pp. 6069–6077.

267. Kim, H., Oh, J., Woo, S. J., Selvapalam, N., Hwang, I., Young, H. K. and Kim, K. (2012). A New Cucurbit[6]uril-Based Ion-Selective Electrode for Acetylcholine with High Selectivity over Choline and Related Quaternary Ammonium Ions. *Supramol. Chem.*, 24, pp. 487–491.

268. Gao, R. H., Chen, L. X., Kai, Chen., Tao, Z. and Xiao, X. (2017). Development of Hydroxylated Cucurbit[*n*]urils, Their Derivatives and Potential Applications. *Coord. Chem. Rev.*, 348, pp. 1–24.

269. Ghale, G., Ramalingam, V., Urbach, A. R. and Nau, W. M. (2011). Determining Protease Substrate Selectivity and Inhibition by Label-Free Supramolecular Tandem Enzyme Assays. *J. Am. Chem. Soc.*, 133, pp. 7528–7535.

270. Ghale, G., Kuhnert, N. and Nau, W. M. (2012). Monitoring Stepwise Proteolytic Degradation of Peptides by Supramolecular Domino Tandem Assays and Mass Spectrometry for Trypsin and Leucine Aminopeptidase. *Nat. Prod. Commun.*, 7, pp. 343–348.

271. Biedermann, F. and Nau, W. M. (2014). Noncovalent Chirality Sensing Ensembles for the Detection and Reaction Monitoring of Amino Acids, Peptides, Proteins, and Aromatic Drugs. *Angew. Chem. Int. Ed.*, 53, pp. 5694–5699.

272. Jang, M., Kim, H., Lee, S., Kim, H. W., Khedkar, J. K., Rhee, Y. M., Hwang, I., Kim, K. and Oh, J. H. (2015). Highly Sensitive and Selective Biosensors Based on Organic Transistors Functionalized with Cucurbit[6] uril Derivatives. *Adv. Funct. Mater.*, 25, pp. 4882–4888.

273. Kim, B., Song, H. S., Jin, H. J., Park, E. J., Lee, S. H., Lee, B. Y., Park, T. H. and Hong, S. (2013). Highly Selective and Sensitive Detection of Neurotransmitters Using Receptor-Modified Single-Walled Carbon Nanotube Sensors. *Nanotechnology*, 24, pp. 285501.

274. Jang, Y., Jang, M., Kim, H., Lee, S. J., Jin, E., Koo, J. Y., Hwang, I.-C., Kim, Y., Ko, Y. H., Hwang, I., Oh, J. H. and Kim, K. (2017). Point-of-Use Detection of Amphetamine-Type Stimulants with Host-Molecule-Functionalized Organic Transistors. *Chem.*, 3, pp. 641–651.

275. Wasserberg, D. and Jonkheijm, P. (2017). Supramolecular Wearable Sensors. *Chem.*, 3, pp. 531–533.

276. del Pozo, M., Hernandez, P., Hernandez, L. and Quintana, C. (2011). The Use of Cucurbit[8]uril Host–Guest Interactions in the Development of an

Electrochemical Sensor: Characterization and Application to Tryptophan Determination. *J. Mater. Chem.*, 21, pp. 13657–13663.

277. Tang, Y., Yang, S., Zhao, Y., You, M., Zhang, F. and He, P. (2015). The Host–Guest Interaction Between Cucurbit[7]uril and Ferrocenemonocarboxylic Acid for Electrochemically Catalytic Determination of Glucose. *Electroanalysis*, 27, 1387–1393.

278. Dantz, D. A., Meschke, C., Buschmann, H.-J. and Schollmeyer, E. (2003). Complexation of Volatile Organic Molecules from the Gas Phase with Cucurbituril and β-Cyclodextrin. *Supramol. Chem.*, 9, pp. 79–83.

279. Buschmann, H. J. and Schollmeyer, E. (1992). Stabilization of Dyes against Hydrolytic Decomposition by the Formation of Inclusion Compounds. *J. Incl. Phenom. Mol. Recognit. Chem.*, 14, pp. 91–99.

280. Karcher, S., Kornmüller, A. and Jekel, M. (2001). Cucurbituril for Water Treatment. Part I: Solubility of Cucurbituril and Sorption of Reactive Dyes. *Water Res.*, 35, pp. 3309–3316.

281. Kornmuller, A., Karcher, S. and Jekel, M. (2001). Cucurbituril for Water Treatment. Part II: Ozonation and Oxidative Regeneration of Cucurbituril. *Water Res.*, 35, pp. 3317–3324.

282. Chen, R., Qiao, H., Liu, Y., Dong, Y., Wang, P., Zhang, Z. and Jin, T. (2014). Adsorption of Methylene Blue from an Aqueous Solution Using a Cucurbituril Polymer. *Environ. Prog. Sustain.*, 34, pp. 512–519.

283. Santos, G., de C. Barros, A. L., de Oliveira, C. A. F., da Luz, L. L., da Silva, F. F., Demets, G. J.-F. and Alves Júnior, S. (2017). New Composites LnBDC@AC and CB[6]@AC: From Design toward Selective Adsorption of Methylene Blue or Methyl Orange. *PLoS One*, 12, p. e0170026.

284. Sousa e Silva, F. C., de Lima, S. M. and Demets, G. J.-F. (2014). Reusable Cucurbit[6]uril-Loaded Poly(urethane) Sponges for Oily Waters Treatment. *RSC Adv.*, 4, pp. 58796–58799.

285. Kim, K., Balaji, R., Oh, D. H., Ko, Y. H. and Jon, S. Y. (2009). Silica Gel Bonded with Cucurbituril. Patent WO 2004072151 A1.

286. Liu, S. M., Xu, L., Wu, C. T. and Feng, Y. Q. (2004). Preparation and Characterization of Perhydroxyl-Cucurbit[6]uril Bonded Silica Stationary Phase for Hydrophilic-Interaction Chromatography. *Talanta*, 64, pp. 929–934.

287. Jeon, Y.-M., Whang, D., Kim, J. and Kim, K. (1996). A Simple Construction of a Rotaxane and Pseudorotaxane: Syntheses and X-Ray Crystal Structures of Cucurbituril Threaded on Substituted Spermine. *Chem. Lett.*, 25, pp. 503–504.

288. Moon, K. and Kaifer, A. E. (2004). Modes of Binding Interaction between Viologen Guests and the Cucurbit[7]uril Host. *Org. Lett.*, 6, pp. 185–188.

289. Whang, D., Heo, J., Kim, C.-A. and Kim, K. (1997). Helical Polyrotaxane: Cucurbituril "Beads" Threaded onto a Helical One-Dimensional Coordination Polymer. *Chem. Commun.*, 33, pp. 2361–2362.

290. Whang, D. and Kim, K. (1997). Polycatenated Two-Dimensional Polyrotaxane Net. *J. Am. Chem. Soc.*, 119, pp. 451–452.

291. Batten, S. R. and Robson, R. (1998). Interpenetrating Nets: Ordered, Periodic Entanglement. *Angew. Chem. Int. Ed.*, 37, pp. 1460–1494.

292. Lee, E., Heo, J. and Kim, K. (2000). A Three-Dimensional Polyrotaxane Network. *Angew. Chem. Int. Ed.*, 39, pp. 2699–2701.

293. Eddaoudi, M., Moler, D. B., Li, H., Chen, B., Reineke, T. M., O'Keeffe, M. and Yaghi, O. M. (2001). Modular Chemistry: Secondary Building Units as a Basis for the Design of Highly Porous and Robust Metal-Organic Carboxylate Frameworks. *Acc. Chem. Res.*, 34, pp. 319–330.

294. Lee, E., Kim, J., Heo, J., Heo, J., Whang, D. and Kim, K. (2001). A Two-Dimensional Polyrotaxane with Large Cavities and Channels: A Novel Approach to Metal-Organic Open-Frameworks by Using Supramolecular Building Blocks. *Angew. Chem. Int. Ed.*, 40, pp. 399–402.

295. Zhu, K. and Loeb, S. J. (2014). Organizing Mechanically Interlocked Molecules to Function Inside Metal–Organic Frameworks. *Top. Curr. Chem.*, 35, 213–551.

296. Leininger, S., Olenyuk, B. and Stang, P. J. (2000). Self-Assembly of Discrete Cyclic Nanostructures Mediated by Transition Metals. *Chem. Rev.*, 100, pp. 853–908.

297. Park, K. M., Kim, S. Y., Heo, J., Whang, D., Sakamoto, S., Yamaguchi, K. and Kim, K. (2002). Designed Self-Assembly of Molecular Necklaces. *J. Am. Chem. Soc.*, 124, pp. 2140–2147.

298. Roh, S.-G., Park, K., Park, G.-J., Sakamoto, S., Yamaguchi, K. and Kim, K. (1999). Synthesis of a Five-Membered Molecular Necklace: A [2 + 2] Approach. *Angew. Chem. Int. Ed.*, 38, pp. 638–641.

299. Buschmann, H.-J., Schollmeyer, E. and Meschke, C. (1999). Polyrotaxanes and Pseudopolyrotaxanes of Polyamides and Cucurbituril. *Polymer*, 20, pp. 945–949.

300. Tuncel, D. and Steinke, J. H. G. (2001). Mainchain Pseudopolyrotaxanes via Post-Threading with Cucurbituril. *Chem. Commun.*, 37, pp. 253–254.

301. Choi, S. W., Lee, J. W., Ko, Y. H. and Kim, K. (2002). Pseudopolyrotaxanes Made to Order: Cucurbituril Threaded on Polyviologen. *Macromolecules*, 35, pp. 3526–3531.

302. Krasia, T. C. and Steinke, J. H. G. (2002). Formation of Oligotriazoles Catalysed by Cucurbituril. *Chem. Commun.*, 38, pp. 22–23.

303. Tuncel, D. and Steinke, J. H. G. (2004). Catalytic Self-Threading: A New Route for the Synthesis of Polyrotaxanes. *Macromolecules*, 37, pp. 288–302.

304. Rekharsky, M. V., Yamamura, H., Kawai, M., Osaka, I., Arakawa, R., Sato, A., Ko, Y. H., Selvapalam, N., Kim, K. and Inoue, Y. (2006). Sequential Formation of a Ternary Complex among Dihexylammonium, Cucurbit[6]uril, and Cyclodextrin with Positive Cooperativity. *Org. Lett.*, 8, pp. 815–818.

305. Yang, C., Young, H. K., Selvapalam, N., Origane, Y., Mori, T., Wada, T., Kim, K. and Inoue, Y. (2007). Dynamic Switching between Single- and Double-Axial Rotaxanes Manipulated by Charge and Bulkiness of Axle Termini. *Org. Lett.*, 9, pp. 4789–4792.

306. Hou, X., Ke, C. and Stoddart, F. J. (2016). Cooperative Capture Synthesis: Yet Another Playground for Copper-Free Click Chemistry. *Chem. Soc. Rev.*, 45, pp. 3766–3780.

307. Ke, C., Strutt, N. L., Li, H., Hou, X., Hartlieb, K. J., McGonigal, P. R., Ma, Z., Iehl, J., Stern, C. L., Cheng, C., Zhu, Z., Vermulen, V. A., Meade, T. J., Botros, Y. Y. and Stoddart, J. F. (2013). Pillar[5]arene as a Co-Factor in Templating Rotaxane Formation. *J. Am. Chem. Soc.*, 135, pp. 17019–17030.

308. Ogoshi, T. (2012). Synthesis of Novel Pillar-Shaped cavitands "Pillar[5] arenes" and Their Application for Supramolecular Materials. *J. Incl. Phenom. Macrocycl. Chem.*, 72, pp. 247–262.

309. Hou, X., Ke, C., Bruns, C. J., McGonigal, P. R., Pettman, R. B. and Stoddart, J. F. (2015). Tunable Solid-State Fluorescent Materials for Supramolecular Encryption. *Nat. Commun.*, 6, p. 6884.

310. Kim, K., Kim, D., Lee, J. W., Ko, Y. H. and Kim, K. (2004). Growth of Poly(pseudorotaxane) on Gold Using Host–Stabilized Charge-Transfer Interaction. *Chem. Commun.*, 40, pp. 848–849.

311. Ko, Y. H., Kim, E., Hwang, I. and Kim, K. (2007). Supramolecular Assemblies Built with Host-Stabilized Charge-Transfer Interactions. *Chem. Commun.*, 43, pp. 1305–1315.

312. Ko, Y. H., Kim, K., Kang, J. K., Chun, H., Lee, J. W., Sakamoto, S., Yamaguchi, K., Fettinger, J. C. and Kim, K. (2004). Designed Self-Assembly of Molecular Necklaces Using Host-Stabilized Charge-Transfer Interactions. *J. Am. Chem. Soc.*, 126, pp. 1932–1933.

313. Li, J., Yu, Y., Luo, L., Li, Y., Wang, P., Cao, L. and Wu, B. (2016). Square[5] Molecular Necklace Formed from Cucurbit[8]uril and Carbazole Derivative. *Tetrahedron Lett.*, 57, pp. 2306–2310.

314. Kim, S. Y., Ko, Y. H., Lee, J. W., Sakamoto, S., Yamaguchi, K. and Kim, K. (2007). Toward High-Generation Rotaxane Dendrimers That Incorporate a

Ring Component on Every Branch: Noncovalent Synthesis of a Dendritic [10]Pseudorotaxane with 13 Molecular Components. *Chem. Asian. J.*, 2, pp. 747–754.

315. Brunsveld, L., Folmer, B. J. B., Meijer, E. W. and Sijbesma, R. P. (2001). Supramolecular Polymers. *Chem. Rev.*, 101, pp. 4071–4098.

316. Del Barrio, J., Horton, P. N., Lairez, D. Lloyd, G. O., Toprakcioglu, C. and Scherman, O. A. (2013). Photocontrol over Cucurbit[8]uril Complexes: Stoichiometry and Supramolecular Polymers. *J. Am. Chem. Soc.*, 135, pp. 11760–11763.

317. Huang, Z., Yang, L., Liu, Y., Wang, Z., Scherman, O. A. and Zhang, X. (2014). Supramolecular Polymerization Promoted and Controlled through Self-Sorting. *Angew. Chem. Int. Ed.* 2014, 53, pp. 5351–5355.

318. Fang, R., Liu, Y., Wang, Z. and Zhang, X. (2013). Water-Soluble Supramolecular Hyperbranched Polymers Based on Host-Enhanced $\pi-\pi$ Interaction. *Polym. Chem.*, 4, pp. 900–903.

319. Pfeffermann, M., Dong, R., Graf, R., Zajaczkowski, W., Gorelik, T., Pisula, W., Narita, A., Müllen, K. and Feng, X. (2015). Free-Standing Monolayer Two-Dimensional Supramolecular Organic Framework with Good Internal Order. *J. Am. Chem. Soc.*, 137, pp. 14525–14532.

320. Ji, Z., Liu, J., Chen, G. and Jiang, M. (2014). A Polymeric Chain Extension Driven by HSCT Interaction. *Polym. Chem.*, 5, pp. 2709–2714.

321. Geng, J., Jiao, D., Rauwald, U. and Scherman, O. A. (2010). An Aqueous Supramolecular Side-Chain Polymer Designed for Molecular Loading. *Aust. J. Chem.*, 63, pp. 627–630.

322. Geng, J., Biedermann, F., Zayed, J. M., Tian, F. and Scherman, O. A. (2011). Supramolecular Glycopolymers in Water: A Reversible Route toward Multivalent Carbohydrate-Lectin Conjugates Using Cucurbit[8]uril. *Macromolecules*, 44 (11), pp. 4276–4281.

323. Moon, K., Grindstaff, J., Sobransingh, D. and Kaifer, A. E. (2004). Cucurbit[8]uril-Mediated Redox-Controlled Self-Assembly of Viologen-Containing Dendrimers. *Angew. Chem. Int. Ed.*, 43, pp. 5496–5499.

324. Zhang, X., Nie, C.-B., Zhou, T.-Y., Qi, Q.-Y., Fu, J., Wang, X.-Z., Dai, L., Chen, Y. and Zhao, X. (2015). The Construction of Single-Layer Two-Dimensional Supramolecular Organic Frameworks in Water through the Self-Assembly of Rigid Vertexes and Flexible Edges. *Polym. Chem.*, 6, pp. 1923–1927.

325. Isaacs, L. (2014). Stimuli Responsive Systems Constructed Using Cucurbit[*n*]uril-Type Molecular Containers. *Acc. Chem. Res.*, 47, pp. 2052–2062.

326. Im Jun, S., Lee, J.W, Sakamoto, S., Yamaguchi, K. and Kim, K. (2000). Rotaxane-Based Molecular Switch with Fluorescence Signaling. *Tetrahedron Lett.*, 41, pp. 471–475.

327. Ooya, T., Inoue, D., Choi, H. S., Kobayashi, Y., Loethen, S., Thompson. D. H., Ko, Y.H., Kim, K. and Yui, N. (2006). pH-Responsive Movement of Cucurbit[7]uril in a Diblock Polypseudorotaxane Containing Dimethyl β-Cyclodextrin and Cucurbit[7]uril. *Org. Lett.* 8, pp. 3159–3162.

328. Sindelar, V., Silvi, S. and Kaifer, A. E. (2006). Switching a Molecular Shuttle On and Off: Simple, pH-Controlled Pseudorotaxanes Based on Cucurbit[7]uril. *Chem. Commun.*, 42, pp. 2185–2187.

329. Lee, J. W., Kim, K. and Kim, K. (2001). A Kinetically Controlled Molecular Switch Based on Bistable [2]rotaxane. *Chem. Commun.*, 37, pp. 1042–1043.

330. Angelos, S., Khashab, N. M., Yang, Y. W., Trabolsi, A., Khatib, H. A., Stoddart, J. F. and Zink, J. I. (2009). pH Clock-Operated Mechanized Nanoparticles. *J. Am. Chem. Soc.*, 131, pp. 12912–12914.

331. Kim, Y., Ko, Y. H., Jung, M., Selvapalam, N. and Kim, K. (2011). A New Photo-Switchable "On–Off" Host–Guest System. *Photochem. Photobiol. Sci.*, 10, pp. 1415–1419.

332. Sun, Y.-L., Yang, B.-J., Zhang, S. X.-A. and Yang, Y.-W. (2012). Cucurbit[7] uril Pseudorotaxane-Based Photoresponsive Supramolecular Nanovalve. *Chem. Eur. J.*, 18, pp. 9212–9216.

333. Tian, F., Jiao, D., Biedermann, F. and Scherman, O. A. (2012). Orthogonal Switching of a Single Supramolecular Complex. *Nat. Commun.*, 5, p. 1207.

334. Baroncini, M., Gao, C., Carboni, V., Credi, A., Previtera, E., Semeraro, M., Venturi, M. and Silvi, S. (2014). Light Control of Stoichiometry and Motion in Pseudorotaxanes Comprising a Cucurbit[7]uril Wheel and an Azobenzene-Bipyridinium Axle. *Chem. Eur. J.*, 20, pp. 10737–10744.

335. Cui, L., Gadde, S., Li, W. and Kaifer, A. E. (2009). Electrochemistry of the Inclusion Complexes Formed Between the Cucurbit[7]uril Host and Several Cationic and Neutral Ferrocene Derivatives. *Langmuir*, 25, pp. 13763–13769.

336. Kaifer, A. E. (2014). Toward Reversible Control of Cucurbit[n]uril Complexes. *Acc. Chem. Res.*, 47, pp. 2160–2167.

337. Jeon, W. S., Ziganshina, A. Y., Lee, J.W., Ko, Y. H., Kang, J. K., Lee, C. and Kim, K. (2003). A [2]Pseudorotaxane-Based Molecular Machine: Reversible Formation of a Molecular Loop Driven by Electrochemical and Photochemical Stimuli. *Angew. Chem. Int. Ed.*, 42, pp. 4097–4100.

338. Rekharsky, M. V., Yamamura, H., Ko, Y. H., Selvapalam, N., Kim, K. and Inoue, Y. (2002). Sequence Recognition and Self-Sorting of a

Dipeptide by Cucurbit[6]uril and Cucurbit[7]uril. *Chem. Commun.*, 38, pp. 2236–2238.

339. Ni, X.-L., Xiao, X., Cong, H., Liang, L.-L., Cheng, K., Cheng, X.-J., Ji, N.-N., Zhu, Q.-J., Xue, S.-F. and Tao, Z. (2013). Cucurbit[*n*]uril-Based Coordination Chemistry: From Simple Coordination Complexes to Novel Poly-Dimensional Coordination Polymers. *Chem. Soc. Rev.*, 42, pp. 9480–9508.

340. Gerasko, O. A., Samsonenko, D. G., Sharonova, A. A., Virovets, A. V., Lipkowski, J. and Fedin, V. P. (2002). Synthesis and Crystal Structure of Supramolecular Adduct of Tetranuclear Uranyl Chloride Aquacomplex with Macrocyclic Cavitand Cucurbituril. *Russ. Chem. Bull.*, 51, pp. 346–349.

341. Heo, J., Kim, S. Y., Whang, D. and Kim, K. (1999). Shape-Induced, Hexagonal, Open Frameworks: Rubidium Ion Complexed Cucurbituril. *Angew. Chem. Int. Ed.*, 38, pp. 641–643.

342. Chen, K., Liang, L. L., Zhang, Y. Q., Zhu, Q. J., Xue, S. F. and Tao, Z. (2011). Novel Supramolecular Assemblies Based on Coordination of Samarium Cation to Cucurbit[5]uril. *Inorg. Chem.*, 50, pp. 7754–7760.

343. Chen, K., Kang, Y. S., Zhao, Y., Yang, J. M., Lu, Y. and Sun, W. Y. (2014). Cucurbit[6]uril-Based Supramolecular Assemblies: Possible Application in Radioactive Cesium Cation Capture. *J. Am. Chem. Soc.*, 136, pp. 16744–16747.

344. Kim, H., Kim, Y., Yoon, M., Lim, S., Park, S., M. Seo, G. and Kim, K. (2010). Highly Selective Carbon Dioxide Sorption in an Organic Molecular Porous Material. *J. Am. Chem. Soc.*, 132, pp. 12200–12202.

345. Tian, J., Liu, J. J., Liu, J. J. and Thallapally, P. K. (2013). Identification of Solid-State Forms of Cucurbit[6]uril for Carbon Dioxide Capture. *Cryst. Eng. Comm.*, 15, pp. 1528–1531.

346. Park, J. H. Suh, K. Rohman, M. R. Hwang, W. Yoon, M. and Kim, K. (2015). Solid Lithium Electrolytes Based on an Organic Molecular Porous Solid. *Chem. Commun.*, 51, pp. 9313–9316.

347. Young, J. F., Nguyen, H. D., Yang, L., Huskens, J., Jonkheijm, P. and Brunsveld, L. (2010). Strong and Reversible Monovalent Supramolecular Protein Immobilization. *ChemBioChem.*, 11, pp. 180–183.

348. Neirynck, P., Brinkmann, J., An, Q., van der Schaft, D. W. J., Milroy, L.-G., Jonkheijm, P. and Brunsveld, L. (2013). Supramolecular Control of Cell Adhesion via Ferrocene-Cucurbit[7]uril Host-Guest Binding on Gold Surfaces. *Chem. Commun.*, 49, pp. 3679–3681.

349. Wasserberg, D., Uhlenheuer, D. A., Neirynck, P., Cabanas-Danés, J., Schenkel, J. H., Ravoo, B. J., An, Q., Huskens, J., Milroy, L. G.,

Brunsveld, L. and Jonkheijm, P. (2014). Immobilization of Ferrocene-Modified SNAP-Fusion Proteins. *Int. J. Mol. Sci.*, 14, pp. 4066–4080.

350. Lee, D. W., Park, K. M., Gong, B., Shetty, D., Khedkar, J. K., Baek, K., Kim, J., Ryu, S. H. and Kim, K. (2015). A Simple Modular Aptasensor Platform Utilizing Cucurbit[7]uril and a Ferrocene Derivative as an Ultrastable Supramolecular Linker. *Chem. Commun.*, 51, pp. 3098–3101.

351. Kim, K., Jeon, W. S., Kang, J. K., Lee, J. W., Jon, S. Y., Kim, T. and Kim, K. A. (2003). Pseudorotaxane on Gold: Formation of Self-Assembled Monolayers, Reversible Dethreading and Rethreading of the Ring, and Ion-Gating Behavior. *Angew. Chem. Int. Ed.*, 42, pp. 2293–2296.

352. Ma, X., Xue, Y., Dai, L., Urbas, A. and Li, Q. (2013). Hydrophilic Cucurbit[7]uril-Pseudorotaxane-Anchored-Monolayer-Protected Gold Nanorods. *Eur. J. Inorg. Chem.*, 2013, pp. 2682–2686.

353. Kang, J.-K., Hwang, I., Young, H. K., Woo, S. J., Kim, H.-J. and Kim, K. (2008). Electrochemically Controllable Reversible Formation of Cucurbit[8]uril-Stabilized Charge-Transfer Complex on Surface. *Supramol. Chem.*, 20, pp. 149–155.

354. Tian, F., Cheng, N., Nouvel, N., Geng, J. and Scherman, O. A. (2010). Site-Selective Immobilization of Colloids on Au Substrates via a Noncovalent Supramolecular "Handcuff". *Langmuir*, 26, pp. 5323–5328.

355. Lan, Y., Wu, Y., Karas, A. and Scherman, O. A. (2014). Photoresponsive Hybrid Raspberry-like Colloids Based on Cucurbit[8]uril Host–Guest Interactions. *Angew. Chem. Int. Ed.*, 53, pp. 2166–2169.

356. Yang, H., An, Q., Zhu, W., Li, W., Jiang, Y., Cui, J., Zhang, X. and Li, G. (2012). A New Strategy for Effective Construction of Protein Stacks by Using Cucurbit[8]uril as a Glue Molecule. *Chem. Commun.*, 48, pp. 10633–10635.

357. Gonzalez-Campo, A., Brasch, M., Uhlenheuer, D. A., Gomez-Casado, A., Yang, L., Brunsveld, L., Huskens, J. and Jonkheijm, P. (2012). Supramolecularly Oriented Immobilization of Proteins Using Cucurbit[8]uril. *Langmuir*, 28, pp. 16364–16371.

358. Kim, J., Kim, Y., Baek, K., Ko, Y. H., Kim, D. and Kim, K. (2008). Direct Force Measurement between Cucurbit[6]uril and Spermine Using Atomic Force Microscopy. *Tetrahedron*, 64, pp. 8389–8393.

359. Zhu, X., Fan, X., Ju, G., Cheng, M., An, Q., Nie, J. and Shi, F. (2013). A Facile Method to Immobilize Cucurbituril on Surfaces through Photocrosslinking with Azido Groups. *Chem. Commun.*, 49, pp. 8093–8095.

360. Gomez-casado, A., Jonkheijm, P. and Huskens, J. (2011). Recognition Properties of Cucurbit[7]uril Self-Assembled Monolayers Studied with Force Spectroscopy. *Langmuir*, 27, pp. 11508–11513.

361. Blanco, E., Quintana, C., Hernández, L. and Hernández, P. (2013). Atomic Force Microscopy Study of New Sensing Platforms: Cucurbit[n]uril (n = 6, 7) on Gold. *Electroanalysis*, 25, pp. 263–268.

362. del Pozo, M., Blanco, E., Fatas, E., Hernandez, P. and Quintana, C. (2012). New Supramolecular Interactions for Electrochemical Sensors Development: Different Cucurbit[8]uril Sensing Platform Designs. *Analyst*, 137, pp. 4302–4308.

363. Qi, L., Tian, H., Shao, H. and Yu, H. (2017). Host–Guest Interaction at Molecular Interfaces: Binding of Cucurbit[7]uril on Ferrocenyl Self-Assembled Monolayers on Gold. *J. Phys. Chem. C*, 121, pp. 7985–7992.

364. Premkumar, T. and Geckeler, K. E. (2006). Nanosized CuO Particles via a Supramolecular Strategy. *Small*, 2, pp. 616–620.

365. Premkumar, T., Lee, Y. and Geckeler, K. E. (2010). Macrocycles as a Tool: A Facile and One-Pot Synthesis of Silver Nanoparticles Using Cucurbituril Designed for Cancer Therapeutics. *Chem. Eur. J.*, 16, pp. 11563–11566.

366. Premkumar, T. and Geckeler, K. E. (2010). Cucurbit[7]uril as a Tool in the Green Synthesis of Gold Nanoparticles. *Chem. Asian J.*, 5, pp. 2468–2476.

367. Lee, T. C. and Scherman, O. A. (2012). A Facile Synthesis of Dynamic Supramolecular Aggregates of Cucurbit[n]uril (n = 5–8) Capped with Gold Nanoparticles in Aqueous Media. *Chem. Eur. J.*, 18, pp. 1628–1633.

368. Barrow, S. J., Kasera, S., Rowland, M. J., Del Barrio, J. and Scherman, O. A. (2015). Cucurbituril-Based Molecular Recognition. *Chem. Rev.*, 115, pp. 12320–12406.

369. Lu, X. and Masson, E. (2011). Formation and Stabilization of Silver Nanoparticles with Cucurbit[n]urils (n = 5–8) and Cucurbituril-Based Pseudorotaxanes in Aqueous Medium. *Langmuir*, 27, pp. 3051–3058.

370. Prodan, E., Radloff, C., Halas, N. J. and Nordlander, P. (2003). A Hybridization Model for the Plasmon Response of Complex Nanostructures. *Science*, 302, pp. 419–422.

371. de Nijs, B., Bowman, R. W., Herrmann, L. O., Benz, F., Barrow, S. J., Mertens, J., Sigle, D. O., Chikkaraddy, R., Eiden, A., Ferrari, A., Scherman, O. A. and Baumberg, J. J. (2015). Unfolding the Contents of Sub-Nm Plasmonic Gaps Using Normalising Plasmon Resonance Spectroscopy. *Faraday Discuss.*, 178, pp. 185–193.

372. Herrmann, L. O., Valev, V. K., Tserkezis, C., Barnard, J. S., Kasera, S., Scherman, O. A., Aizpurua, J. and Baumberg, J. J. (2014). Threading Plasmonic Nanoparticle Strings with Light. *Nat. Commun.*, 5, p. 4568.

373. Yeh, Y.-C., Rana, S., Mout, R., Yan, B., Alfonso, F. S. and Rotello, V. M. (2014). Supramolecular Tailoring of Protein-Nanoparticle Interactions Using Cucurbituril Mediators. *Chem. Commun.*, 50, pp. 5565–5568.

374. Jones, S. T., Zayed, J. M. and Scherman, O. A. (2013). Supramolecular Alignment of Gold Nanorods via Cucurbit[8]uril Ternary Complex Formation. *Nanoscale*, 5, pp. 5299–5302.

375. Cao, M., Lin, J., Yang, H. and Cao, R. (2010). Facile Synthesis of Palladium Nanoparticles with High Chemical Activity Using Cucurbit[6]uril as Protecting Agent. *Chem. Commun.*, 46, pp. 5088–5090.

376. Cao, M., Wu, D., Gao, S. and Cao, R. (2012). Platinum Nanoparticles Stabilized by Cucurbit[6]uril with Enhanced Catalytic Activity and Excellent Poisoning Tolerance for Methanol Electrooxidation. *Chem. Eur. J.*, 18, pp. 12978–12985.

377. Lanterna, A., Pino, E., Doménech-Carbó, A., González-Béjar, M. and Pérez-Prieto, J. (2014). Enhanced Catalytic Electrochemical Reduction of Dissolved Oxygen with Ultraclean Cucurbituril[7]-Capped Gold Nanoparticles. *Nanoscale*, 6, pp. 9550–9553.

378. Premkumar, T. and Geckeler, K. E. (2014). Synthesis of Honeycomb-Like Palladium Nanostructures by Using Cucurbit[7]uril and Their Catalytic Activities for Reduction of 4-Nitrophenol. *Mater. Chem. Phys.*, 148, pp. 772–777.

379. Blanco, E., Esteve-Adell, I., Atienzar, P., Casas, J. A., Hernandez, P. and Quintana, C. (2016). Cucurbit[7]uril-Stabilized Gold Nanoparticles as Catalysts of the Nitro Compound Reduction Reaction. *RSC Adv.*, 6, pp. 86309–86315.

380. Li, H., Lü, J., Lin, J., Huang, Y., Cao, M. and Cao, R. (2013). Crystalline Hybrid Solid Materials of Palladium and Decamethylcucurbit[5]uril as Recoverable Precatalysts for Heck Cross-Coupling Reactions. *Chem. Eur. J.*, 19, pp. 15661–15668.

381. Li, H. F., Lü, J., Lin, J. X. and Cao, R. (2014). Monodispersed Ag Nanoparticles as Catalyst: Preparation Based on Crystalline Supramolecular Hybrid of Decamethylcucurbit[5]uril and Silver Ions. *Inorg. Chem.*, 53, pp. 5692–5697.

382. Blackie, E. J., Le Ru, E. C. and Etchegoin, P. G. (2009). Single-Molecule Surface-Enhanced Raman Spectroscopy of Nonresonant Molecules. *J. Am. Chem. Soc.*, 131, pp. 14466–14472.

383. Nie, S. and Emory, S. R. (1997). Probing Single Molecules and Single Nanoparticles by Surface Enhanced Raman Scattering. *Science*, 275, pp. 1102–1106.

384. Mahajan, S., Lee, T.-C., Biedermann, F., Hugall, J. T., Baumberg, J. J. and Scherman, O. A. (2010). Raman and SERS Spectroscopy of Cucurbit[*n*] urils. *Phys. Chem. Chem. Phys.*, 12, pp. 10429–10433.

385. Tao, C.-A., An, Q., Zhu, W., Yang, H. W., Li, W. H., Lin, C., Xu, D. and Li, G. (2011). Cucurbit[*n*]urils as a SERS Hot-Spot Nanocontainer through Bridging Gold Nanoparticles. *Chem. Commun.*, 47, pp. 9867–9869.

386. El-barghouthi, M. I., Assaf, K. I. and Rawashdeh, A. M. M. (2010). Molecular Dynamics of Methyl Viologen-Cucurbit[*n*]uril Complexes in Aqueous Solution. *J. Chem. Theory Comput.*, 6, pp. 984–992.

387. Kasera, S., Biedermann, F., Baumberg, J. J., Scherman, O. A. and Mahajan, S. (2012). Quantitative SERS Using the Sequestration of Small Molecules Inside Precise Plasmonic Nanoconstructs. *Nano Lett.*, 12, pp. 5924–5928.

388. Chen, Y., Klimczak, A., Galoppini, E. and Lockard, J. V. (2013). Structural Interrogation of a Cucurbit[7]uril-Ferrocene Host–Guest Complex in the Solid State: A Raman Spectroscopy Study. *RSC Adv.*, 3, pp. 1354–1358.

389. Roldán, M. L., Sanchez-Cortes, S., García-Ramos, J. V. and Domingo, C. (2012). Cucurbit[8]uril-Stabilized Charge Transfer Complexes with Diquat Driven by pH: A SERS Study. *Phys. Chem. Chem. Phys.*, 14, pp. 4935–4941.

390. Taylor, R. W., Coulston, R. J., Biedermann, F., Mahajan, S., Baumberg, J. J. and Scherman, O. A. (2013). *In Situ* SERS Monitoring of Photochemistry within a Nanojunction Reactor. *Nano Lett.*, 13, pp. 5985–5990.

391. Chikkaraddy, R., de Nijs, B., Benz, F., Barrow, S. J., Scherman, O. A., Rosta, E., Demetriadou, A., Fox, P., Hess, O. and Baumberg, J. J. (2016). Single-Molecule Strong Coupling at Room Temperature in Plasmonic Nanocavities. *Nature*, 535, pp.127–130.

392. Whitesides, G. M. and Grzybowski, B. (2002). Self-Assembly at All Scales. *Science*, 295, pp. 2418–2421.

393. Lehn, J.-M. (2007). From Supramolecular Chemistry towards Constitutional Dynamic Chemistry and Adaptive Chemistry. *Chem. Soc. Rev.* 2007, 36, pp. 151–160.

394. Baek, K., Hwang, I., Roy, I., Shetty, D. and Kim, K. (2015). Self-Assembly of Nanostructured Materials through Irreversible Covalent Bond Formation. *Acc. Chem. Res.*, 48, pp. 2221–2229.

395. Kim, D., Kim, E., Lee, J., Hong, S., Sung, W., Lim, N., Park, C. G. and Kim, K. (2010). Direct Synthesis of Polymer Nanocapsules: Self-Assembly of Polymer Hollow Spheres through Irreversible Covalent Bond Formation. *J. Am. Chem. Soc.*, 132, pp. 9908–9919.

396. Kim, S., Yun, G., Khan, S., Kim, J., Murray, J., Lee, M., Kim, J., Lee, G., Kim, S., Shetty, D., Kang, J. H., Kim, J. Y., Park, K. M. and kim, K. (2017). Cucurbit[6]uril-Based Polymer Nanocapsules for Multimodal *In Vivo* Imaging. *Mater. Horizons*, 4, pp. 450–455.

397. Yun, G., Hassan, Z., Lee, J., Kim, J., Lee, N. S., Kim, N. H., Baek, K., Hwang, I., Park, C. G. and Kim, K. (2014). Highly Stable, Water-Dispersible Metal-Nanoparticle-Decorated Polymer Nanocapsules and Their Catalytic Applications. *Angew. Chem. Int. Ed.*, 53, pp. 6414–6418.

398. Sakamoto, J., van Heijst, J., Lukin, O. and Schlüter, A. D. (2009). Two-Dimensional Polymers: Just a Dream of Synthetic Chemists? *Angew. Chem. Int. Ed.*, 48, pp. 1030–1069.

399. Baek, K., Xu, D., Murray, J., Kim, S. and Kim, K. (2016). Permselective 2D-Polymer-Based Membrane Tuneable by Host–Guest Chemistry. *Chem. Commun.*, 52, pp. 9676–9678.

400. Liu, J., Lan, Y., Yu, Z., Tan, C. S. Y., Parker, R. M., Abell, C. and Scherman, O. A. (2017). Cucurbit[*n*]uril-Based Microcapsules Self-Assembled within Microfluidic Droplets: A Versatile Approach for Supramolecular Architectures and Materials. *Acc. Chem. Res.*, 50, pp. 208–217.

401. Zheng, Y., Yu, Z., Parker, R. M., Wu, Y., Abell, C. and Scherman, O. A. (2014). Interfacial Assembly of Dendritic Microcapsules with Host–Guest Chemistry. *Nat. Commun.*, 5, pp. 5772.

402. Yang, H., Tan, Y. and Wang, Y. (2009). Fabrication and Properties of Cucurbit[6]uril Induced Thermo-Responsive Supramolecular Hydrogels. *Soft Matter*, 5, pp. 3511–3516.

403. Jung, H., Park, J. S., Yeom, J., Selvapalam, N., Park, K. M., Oh, K., Yang, J. A., Park, K. H., Hahn, S. K. and Kim, K. (2014). 3D Tissue Engineered Supramolecular Hydrogels for Controlled Chondrogenesis of Human Mesenchymal Stem Cells. *Biomacromolecules*, 15, pp. 707–714.

404. Hwang, B. W., Kim, S. J., Park, K. M., Kim, H., Yeom, J., Yang, J. A., Jeong, H., Jung, H., Kim, K., Sung, Y. C. and Hahn, S. K. (2015). Genetically Engineered Mesenchymal Stem Cell Therapy Using Self-Assembling Supramolecular Hydrogels. *J. Control. Release*, 220, pp. 121–129.

405. Yeom, J., Kim, S. J., Jung, H., Namkoong, H., Yang, J., Hwang, B. W., Oh, K., Kim, K., Sung, Y. C. and Hahn, S. K. (2015). Supramolecular Hydrogels

for Long-Term Bioengineered Stem Cell Therapy. *Adv. Healthc. Mater.*, 4, pp. 237–244.

406. Chen, H., Hou, S., Ma, H., Li, X. and Tan, Y. (2016). Controlled Gelation Kinetics of Cucurbit[7]uril-Adamantane Cross-Linked Supramolecular Hydrogels with Competing Guest Molecules. *Sci. Rep.*, 6, pp. 20722.

407. Appel E. A. and Loh, X. J. (2012). Ultrahigh-Water-Content Supramolecular Hydrogels Exhibiting Multistimuli Responsiveness. *J. Am. Chem. Soc.*, 134, pp. 11767–11773.

408. Li, C., Rowland, M. J., Shao, Y., Cao, T., Chen, C., Jia, H., Zhou, X., Yang, Z., Scherman, O. A. and Liu, D. (2015). Responsive Double Network Hydrogels of Interpenetrating DNA and CB[8] Host–Guest Supramolecular Systems. *Adv. Mater.*, 27, pp. 3298–3304.

409. Walsh, Z., Janeček, E.-R., Hodgkinson, J. T., Sedlmair, J., Koutsioubas, A., Spring, D. R., Welch, M., Hirschmugl, C. J., Toprakcioglu, C., Nitschke, J. R., Jones, M. and Scherman, O. A. (2014). Multifunctional Supramolecular Polymer Networks as Next-Generation Consolidants for Archaeological Wood Conservation. *Proc. Natl. Acad. Sci.*, 111, pp. 17743–17748.

410. Appel, E. A., Forster, R. A., Koutsioubas, A., Toprakcioglu, C. and Scherman, O. A. (2014). Activation Energies Control the Macroscopic Properties of Physically Cross-Linked Materials. *Angew. Chem. Int. Ed.*, 53, pp. 10038–10043.

411. Appel, E. A. Forster, R. A. Rowland, M. J. and Scherman, O. A. (2014). The Control of Cargo Release from Physically Crosslinked Hydrogels by Crosslink Dynamics. *Biomaterials*, 35, pp. 9897–9903.

412. Biedermann, F., Ross, I. and Scherman, O. A. (2014). Host–Guest Accelerated Photodimerisation of Anthracene-Labeled Macromolecules in Water. *Polym. Chem.*, 5, pp. 5375–5382.

413. Liu, J., Tan, C. S. Y., Yu, Z., Lan, Y., Abell, C. and Scherman, O. A. (2017). Biomimetic Supramolecular Polymer Networks Exhibiting both Toughness and Self-Recovery. *Adv. Mater.*, 29, p. 1604951.

414. Liu, J., Tan, C. S. Y., Yu, Z., Li, N., Abell, C. and Scherman, O. A. (2017). Tough Supramolecular Polymer Networks with Extreme Stretchability and Fast Room-Temperature Self-Healing. *Adv. Mater.*, 29, p. 1605325.

415. Park, K. M., Roh, H., Sung, G., Murray, J. and Kim, K. (2017). Self-Healable Supramolecular Hydrogel Formed by Nor-Seco-Cucurbit[10] uril as a Supramolecular Crosslinker. *Chem. Asian J.* 2017, 12, pp. 1461–1464.

416. Buschmann, H. J., Schollmeyer, E. and Mutihac, L. (2003). The Formation of Amino Acid and Dipeptide Complexes with α-Cyclodextrin and

Cucurbit[6]uril in Aqueous Solutions Studied by Titration Calorimetry. *Thermochim. Acta*, 399, pp. 203–208.

417. Lagona, J., Wagner, B. D. and Isaacs, L. (2006). Molecular-Recognition Properties of a Water-Soluble Cucurbit[6]uril Analogue. *J. Org. Chem.*, 71, pp. 1181–1190.

418. Choi, T. S., Lee, H. H., Ko, Y. H., Jeong, K. S., Kim, K. and Kim, H. I. (2017). Nanoscale Control of Amyloid Self-Assembly Using Protein Phase Transfer by Host–Guest Chemistry. *Sci. Rep.*, 7, p. 5710.

419. Urbach, A. R. and Ramalingam, V. (2011). Molecular Recognition of Amino Acids, Peptides, and Proteins by Cucurbit[*n*]uril Receptors. *Isr. J. Chem.*, 51, pp. 664–678.

420. Lee, H. H., Choi, T. S., Lee, S. J. C., Lee, J. W., Park, J., Ko, Y. H., Kim, W. J., Kim, K. and Kim, H. I. (2014). Supramolecular Inhibition of Amyloid Fibrillation by Cucurbit[7]uril. *Angew. Chem. Int. Ed.*, 53, pp. 7461–7465.

421. Logsdon, L. A., Schardon, C. L., Ramalingam, V., Kwee, S. K. and Urbach, A. R. (2011). Nanomolar Binding of Peptides Containing Noncanonical Amino Acids by a Synthetic Receptor. *J. Am. Chem. Soc.*, 133, pp. 17087–17092.

422. Rajgariah, P. and Urbach, A. R. (2008). Scope of Amino Acid Recognition by Cucurbit[8]uril. *J. Incl. Phenom. Macrocycl. Chem.*, 62, pp. 251–254.

423. Ling, Y., Wang, W. and Kaifer, A. E. (2007). A New Cucurbit[8]uril-Based Fluorescent Receptor for Indole Derivatives. *Chem. Commun.*, 43, pp. 610–612.

424. Tian, F., Cziferszky, M., Jiao, D., Wahlström, K., Geng, J. and Scherman, O. A. (2011). Peptide Separation through a CB[8]-Mediated Supramolecular Trap-and-Release Process. *Langmuir*, 27, pp. 1387–1390.

425. Biedermann, F., Rauwald, U., Cziferszky, M., Williams, K. A., Gann, L. D., Guo, B. Y., Urbach, A. R., Bielawski, C. W. and Scherman, O. A. (2011). Benzobis(imidazolium)-Cucurbit[8]uril Complexes for Binding and Sensing Aromatic Compounds in Aqueous Solution. *Chem. Eur. J.*, 16, pp. 13716–13722.

426. Smith, L. C., Leach, D. G., Blaylock, B. E., Ali, O. A. and Urbach, A. R. (2015). Sequence-Specific, Nanomolar Peptide Binding via Cucurbit[8] uril-Induced Folding and Inclusion of Neighboring Side Chains. *J. Am. Chem. Soc.*, 137, pp. 3663–3669.

427. Sonzini, S., Ryan, S. T. J. and Scherman, O. A. (2013). Supramolecular Dimerisation of Middle-Chain Phe Pentapeptides via CB[8] Host–Guest Homoternary Complex Formation. *Chem. Commun.*, 49, pp. 8779–8781.

428. Biedermann, F., Rauwald, U., Zayed, J. M. and Scherman, O. A. (2011). A Supramolecular Route for Reversible Protein–Polymer Conjugation. *Chem. Sci.*, 2, pp. 279–286.
429. Nguyen, H. D., Dang, D. T., Van Dongen, J. L. J. and Brunsveld, L. (2010). Protein Dimerization Induced by Supramolecular Interactions with Cucurbit[8] uril. *Angew. Chem. Int. Ed.*, 49, pp. 895–898.
430. Dang, D. T., Schill, J. and Brunsveld, L. (2012). Cucurbit[8]uril-Mediated Protein Homotetramerization. *Chem. Sci.*, 3, p. 2679.
431. Ramaekers, M., Wijnands, S. P. W., Van Dongen, J. L. J., Brunsveld, L. and Dankers, P. Y. W. (2015). Cucurbit[8]uril Templated Supramolecular Ring Structure Formation and Protein Assembly Modulation. *Chem. Commun.*, 51, pp. 3147–3150.
432. Bosmans, R. P. G., Briels, J. M., Milroy, L. G., de Greef, T. F. A., Merkx, M. and Brunsveld, L. (2016). Supramolecular Control Over Split-Luciferase Complementation. *Angew. Chem. Int. Ed.*, 55, pp. 8899–8903.
433. Hou, C., Li, J., Zhao, L., Zhang, W., Luo, Q., Dong, Z., Xu, J. and Liu, J. (2013). Construction of Protein Nanowires through Cucurbit[8]uril-Based Highly Specific Host-Guest Interactions: An Approach to the Assembly of Functional Proteins. *Angew. Chem. Int. Ed.*, 52, pp. 5590–5593.
434. Si, C., Li, J., Luo, Q., Hou, C., Pan, T., Li, H. and Liu, J. (2016). An Ion Signal Responsive Dynamic Protein Nano–Spring Constructed by High Ordered Host–Guest Recognition. *Chem. Commun.*, 52, pp. 2924–2927.
435. Huang, Z., Fang, Y., Luo, Q., Liu, S., An, G., Hou, C., Lang, C., Xu, J., Dong, Z. and Liu, J. (2016). Construction of Supramolecular Polymer by Enzyme-Triggered Covalent Condensation of CB[8]-FGG-Based Supramonomer. *Chem. Commun.*, 52, pp. 2083–2086.
436. Saleh, N., Ghosh, I. and Nau, W. M. (2013). *Supramolecular Systems in Biomedical Fields,* Ed. Schneider, H. J., Chapter 7, Cucurbiturils in Drug Delivery And For Biomedical Applications (Royal Society of Chemistry, London).
437. Oun, R., Floriano, R. S., Isaacs, L., Rowan, E. G. and Wheate, N. J. (2014). The *Ex Vivo* Neurotoxic, Myotoxic and Cardiotoxic Activity of Cucurbituril-Based Macrocyclic Drug Delivery Vehicles. *Toxicol. Res.*, 3, pp. 447–455.
438. Chen, H., Chan, J. Y. W., Yang, X., Wyman, I. W., Bardelang, D., Macartney, D. H., Lee, S. M. Y. and Wang, R. (2015). Developmental and Organ-Specific Toxicity of Cucurbit[7]uril: *In Vivo* Study on Zebrafish Models. *RSC Adv.*, 5, pp. 30067–30074.

439. Wheate, N. J., Day, A. I., Blanch, R. J., Arnold, A. P., Cullinane, C. and Collins, J. G. (2004). Multi-Nuclear Platinum Complexes Encapsulated in Cucurbit[n]uril as an Approach to Reduce Toxicity in Cancer Treatment. *Chem. Commun.*, 40, pp. 1424–1425.

440. Wheate, N. J. (2008). Improving Platinum(II)-Based Anticancer Drug Delivery Using Cucurbit[n]urils. *J. Inorg. Biochem.*, 102, pp. 2060–2066.

441. Jeon, Y. J., Kim, S.-Y., Ho Ko, Y., Sakamoto, S., Yamaguchi, K. and Kim, K. (2005). Novel Molecular Drug Carrier: Encapsulation of Oxaliplatin in Cucurbit[7]uril and Its Effects on Stability and Reactivity of the Drug. *Org. Biomol. Chem.*, 3, pp. 2122–2125.

442. Appel, E. A., Rowland, M. J., Loh, X. J., Scherman, O. A., Heywood, R. M. and Watts, C. (2012). Enhanced Stability and Activity of Temozolomide in Primary Glioblastoma Multiforme Cells with Cucurbit[n]uril. *Chem. Commun.*, 48, pp. 9843–9845.

443. Walker, S., Kaur, R., Mcinnes, F. J. and Wheate, N. J. (2010). Synthesis, Processing and Solid State Excipient Interactions of Cucurbit[6]uril and Its Formulation into Tablets for Oral Drug Delivery. *Mol. Pharm.*, 7, pp. 2166–2172.

444. Seif, M., Impelido, M. L., Apps, M. G. and Wheate, N. J. (2014). Topical Cream-Based Dosage Forms of the Macrocyclic Drug Delivery Vehicle Cucurbit[6]uril. *PLoS One*, 9, p. e85361.

445. McInnes, F. J., Anthony, N. G., Kennedy, A. R. and Wheate, N. J. (2010). Solid State Stabilisation of the Orally Delivered Drugs Atenolol, Glibenclamide, Memantine and Paracetamol through Their Complexation with Cucurbit[7]uril. *Org. Biomol. Chem.*, 8, pp. 765–773.

446. Wheate, N. J. and Limantoro, C. (2016). Cucurbit[n]urils as Excipients in Pharmaceutical Dosage Forms. *Supramol. Chem.*, 28, pp. 849–856.

447. Zhao, Y., Buck, D. P., Morris, D. L., Pourgholami, M. H., Day, A. I. and Collins, J. G. (2008). Solubilisation and Cytotoxicity of Albendazole Encapsulated in Cucurbit[n]uril. *Org. Biomol. Chem.*, 6, p. 4509.

448. Shaikh, M., Mohanty, J., Bhasikuttan, A. C., Uzunova, V. D., Nau, W. M. and Pal, H. (2008). Salt-Induced Guest Relocation from a Macrocyclic Cavity into a Biomolecular Pocket: Interplay between Cucurbit[7]uril and Albumin. *Chem. Commun.*, 44, pp. 3681–3683.

449. Carvalho, C. P., Uzunova, V. D., Da Silva, J. P., Nau, W. M. and Pischel, U. (2011). A Photoinduced pH Jump Applied to Drug Release from Cucurbit[7] uril. *Chem. Commun.*, 47, pp. 8793–8795.

450. Vázquez, J., Romero, M. A., Dsouza, R. N. and Pischel, U. (2016). Phototriggered Release of Amine from a Cucurbituril Macrocycle. *Chem. Commun.*, 52, pp. 6245–6248.

451. Kim, K., Selvapalam, N., Ko, Y. H,. Park, K. M., Kim, D. and Kim, J. (2007). Functionalized Cucurbiturils and Their Applications. *Chem. Soc. Rev.*, 36, pp. 267–279.

452. Lee, H. K., Park, K. M., Jeon, Y. J., Kim, D., Oh, D. H., Kim, H. S., Park, C. K. and Kim, K. (2005). Vesicle Formed by Amphiphilc Cucurbit[6]uril: Versatile, Noncovalent Modification of the Vesicle Surface, and Multivalent Binding of Sugar-Decorated Vesicles to Lectin. *J. Am. Chem. Soc.*, 127, pp. 5006–5007.

453. Park, K. M., Lee, D. W., Sarkar, B., Jung, H., Kim, J., Ko, Y. H., Lee, K. E., Jeon, H. and Kim, K. (2010). Reduction-Sensitive, Robust Vesicles with a Noncovalently Modifiable Surface as a Multifunctional Drug–Delivery Platform. *Small*, 6, pp. 1430–1441.

454. Park, K. M., Suh, K., Jung, H., Lee, D.-W., Ahn, Y., Kim, J., Baek, K. and Kim, K. (2009). Cucurbituril-Based Nanoparticles: A New Efficient Vehicle for Targeted Intracellular Delivery of Hydrophobic Drugs. *Chem. Commun.*, 45, pp. 71–73.

455. Jung, H., Park, K. M., Yang, J. A., Oh, E. J., Lee, D. W., Park, K., Ryu, S. H., Hahn, S. K. and Kim, K. (2011). Theranostic Systems Assembled *In Situ* on Demand by Host–Guest Chemistry. *Biomaterials*, 32, pp. 7687–7694.

456. Kim, E., Lee, J., Kim, D., Lee, K. E., Han, S. S., Lim, N., Kang, J., Park, C. G. and Kim, K. (2009). Solvent-Responsive Polymer Nanocapsules with Controlled Permeability: Encapsulation and Release of a Fluorescent Dye by Swelling and Deswelling. *Chem. Commun.*, 45, pp. 1472–1474.

457. Kim, E., Kim, D., Jung, H., Lee, J., Paul, S., Selvapalam, N., Yang, Y., Lim, N., Park, C. G. and Kim, K. (2010). Facile, Template-Free Synthesis of Stimuli-Responsive Polymer Nanocapsules for Targeted Drug Delivery. *Angew. Chem. Int. Ed.*, 49, pp. 4405–4408.

458. Jiao, D., Geng, J., Loh, X. J., Das, D., Lee, T. C. and Scherman, O. A. (2012). Supramolecular Peptide Amphiphile Vesicles through Host–Guest Complexation. *Angew. Chem. Int. Ed.*, 51, pp. 9633–9637.

459. Loh, X. J., Del Barrio, J., Toh, P. P. C., Lee, T. C., Jiao, D., Rauwald, U., Appel, E. A. and Scherman, O. A. (2012). Triply Triggered Doxorubicin Release from Supramolecular Nanocontainers. *Biomacromolecules*, 13, pp. 84–91.

460. Loh, X. J., Tsai, M.-H., Barrio, J., del Appel, E. A., Lee, T.-C. and Scherman, O. A. (2012). Triggered Insulin Release Studies of Triply Responsive Supramolecular Micelles. *Polym. Chem.*, 3, pp. 3180–3188.

461. Chen, C.-J., Li, D.-D., Wang, H.-B., Zhao, J. and Ji, J. (2013). Fabrication of Dual-Responsive Micelles Based on the Supramolecular Interaction of Cucurbit[8]uril. *Polym. Chem.*, 4, pp. 242–245.

462. Ambrogio, M. W., Thomas, C. R., Zhao, Y. L., Zink, J. I. and Stoddart, J. F. (2011). Mechanized Silica Nanoparticles: A New Frontier in Theranostic Nanomedicine. *Acc. Chem. Res.*, 44, pp. 903–913.

463. Thomas, C. R., Ferris, D. P., Lee, J. H., Choi, E., Cho, M. H., Kim, E. S., Stoddart, J. F., Shin, J. S., Cheon, J. and Zink, J. I. (2010). Noninvasive Remote-Controlled Release of Drug Molecules *In Vitro* Using Magnetic Actuation of Mechanized Nanoparticles. *J. Am. Chem. Soc.*, 132, pp. 10623–10625.

464. Jeon, Y. J., Kim, H., Jon, S., Selvapalam, N., Dong, H. O., Seo, I., Park, C. S., Seung, R. J., Koh, D. S. and Kim, K. (2004). Artificial Ion Channel Formed by Cucurbit[n]uril Derivatives with a Carbonyl Group Fringed Portal Reminiscent of the Selectivity Filter of K^+ Channels. *J. Am. Chem. Soc.*, 126, pp. 15944–15945.

465. Kim, B. S., Ko, H., Kim, Y., Lee, J., Selvapalam, N., Lee, C. and Kim, K. (2008). Water Soluble Cucurbit[6]uril Derivative as a Potential Xe Carrier for Xe NMR-Based Biosensors. *Chem. Commun.*, 44, pp. 2756–2758.

466. Wang, Y. and Dmochowski, I. J. (2015). Cucurbit[6]uril is an Ultrasensitive ^{129}Xe NMR Contrast Agent. *Chem. Commun.*, 51, pp. 8982–8985.

467. Hane, F. T., Li, T., Smylie, P., Pellizzari, R. M., Plata, J. A., DeBoef, B. and Albert, M. S. (2016). *In Vivo* Detection of a Hyperpolarized Xenon Magnetic Resonance Molecular Imaging Contrast Agent. *Sci. Rep.*, 7, p. 41027.

468. Green, N. M. (1975). Avidin. *Adv. Protein. Chem.*, 29, pp. 85–133.

469. Piran, U. and Riordan, W. J. (1990). Dissociation Rate Constant of the Biotin-Streptavidin Complex. *J. Immunol. Methods*, 133, pp. 141–143.

470. Bae, Y.-I., Hwang, I., Kim, I., Kim, K. and Park, J. W. (2017). Force Measurement for the Interaction between Cucurbit[7]uril and Mica and Self-Assembled Monolayer in the Presence of Zn^{2+} Studied with Atomic Force Microscopy. *Langmuir*, 33, pp. 11884–11892.

471. Murray, J., Sim, J., Oh, K., Sung, G., Lee, A., Shrinidhi, A., Thirunarayanan, A., Shetty, D. and Kim, K. (2017). Enrichment of Specifically Labeled Proteins by an Immobilized Host Molecule. *Angew. Chem. Int. Ed.*, 56, pp. 2395–2398.

472. Li, W., Bockus, A. T., Vinciguerra, B., Isaacs, L. D. and Urbach, A. R. (2016). Predictive Recognition of Native Proteins by Cucurbit[7]uril in a Complex Mixture. *Chem. Commun.*, 52, pp. 8537–8540.

473. Lee, B. P., Messersmith, P. B., Israelachvili, J. N. and Waite, J. H. (2011). Mussel-inspired Adhesives and Coatings. *Annu. Rev. Mater. Res.*, 41, pp. 99–132.

474. Guo, J., Yuan, C., Guo, M., Wang, L. and Yan, F. (2014). Flexible and Voltage-Switchable Polymer Velcro Constructed Using Host–Guest Recognition between Poly(Ionic Liquid) Strips. *Chem. Sci.*, 5, p. 3261.

475. Kim, C., Yan, B., Kim, C. S., Kim, S. T., Park, M., Zhu, Z., Duncan, B., Creran, B. and Rotello, V. M. (2015). Regulating Exocytosis of Nanoparticles via Host–Guest Chemistry. *Org. Biomol. Chem.*, 13, pp. 2474–2479.
476. Ghosh, S. and Isaacs, L. (2010). Biological Catalysis Regulated by Cucurbit[7] uril Molecular Containers. *J. Am. Chem. Soc.*, 132, pp. 4445–4454.
477. Park, K. M., Murray, J. and Kim, K. (2017). Ultrastable Artificial Binding Pairs as a Supramolecular Latching System: A Next Generation Chemical Tool for Proteomics. *Acc. Chem. Res.*, 50, pp. 644–646.
478. Kaabel, S., Adamson, J., Topić, F., Kiesilä, A., Kalenius, E., Öeren, M., Reimund, M., Prigorchenko, E., Lõokene, A., Reich, H. J., Rissanen, K. and Aav, R. (2016). Chiral Hemicucurbit[8]uril as an Anion Receptor: Selectivity to Size, Shape and Charge Distribution. *Chem. Sci.*, 8, pp. 2184–2190.
479. Hashidzume, A., Hiroyasu, Y. and Harada, A. (2014). Cyclodextrin-Based Molecular Machines. *Top. Curr. Chem.*, 354, pp. 471–110.
480. Mock, W. L. (1995). Cucurbituril. *Top. Curr. Chem.*, 175, pp. 1–24.
481. Kim, K. (1999). *Perspectives in Supramolecular Chemistry: Transition Meats in Supramolecular Chemistry,* Ed. J. P. Sauvage, (Wiley, New York), pp. 371–402.
482. Lee, J. W. and Kim, K. Rotaxane Dendrimers. *Top. Curr. Chem.*, 228, pp. 111–140.
483. Lagona, J., Mukhopadhyay, P., Chakrabarti, S. and Isaacs, L. (2005). The Cucurbit[n]uril Family. *Angew. Chem. Int. Ed.*, 44, pp. 4844–4870.
484. Korner, A. L. and Nau, W. M. (2007). Cucurbituril Encapsualtion of Fluorescent Dyes. *Supramol. Chem.*, 19, pp. 55–66.
485. Walker, S., Oun, R., McInnes, F. J. and Wheate, N. J. (2011). The Potential of Cucurbit[n]urils in Drug Delivery. *Isr. J. Chem.*, 51, pp. 616–624.
486. Gadde, S. and Kaifer, A. E. (2011). Cucurbituril Complexes of Redox Active Guests. *Curr. Org. Chem.*, 15, pp. 27–38.
487. Kaifer, A. E., Li, W. and Yi, S. (2011). Cucurbiturils as Versatile Receptors for Redox Active Substrates. *Isr. J. Chem.*, 51, pp. 496–505.
488. Ko, Y. H., Hwang, I., Lee, D. W. and Kim, K. (2011). Ultrastable Host–Guest Complexes and Their Applications. *Isr. J. Chem.*, 51, pp. 506–514.
489. Parvari, G., Reany, O. and Keinan, E. (2011). Applicable Properties of Cucurbiturils. *Isr. J. Chem.*, 51, pp. 646–663.
490. Das, D. and Scherman, O. A. (2011). Cucurbituril: At the Interface of Small Molecule Host–Guest Chemistry and Dynamic Aggregates. *Isr. J. Chem.*, 51, pp. 537–550.
491. Chen, Y., Zhang, Y.-M. and Liu, Y. (2011). Molecular Selective Binding and Nanofabrication of Cucurbituril/Cyclodextrin Pairs. *Isr. J. Chem.*, 51, pp. 515–524.

492. H. J. Buschmann. (2011). From Small Cucurbituril Complexes to Large Ordered Networks. *Isr. J. Chem.*, 51, pp. 533–536.

493. Stancl, M., Svec, J. and Sindelar, V. (2011). Novel Supramolecular Hosts Based on Linear and Cyclic Oligomers of Glycoluril. *Isr. J. Chem.*, 51, pp. 592–599.

494. Biczók, L., Wintgens, V., Miskolczy, Z. and Megyesi, M. (2011). Fluorescence Response of Alkaloids and DAPI on Inclusion in Cucurbit[7] uril: Utilization for the Study of the Encapsulation of Ionic Liquid Cations. *Isr. J. Chem.*, 51, pp. 625–633.

495. Bhasikuttan, A. C., Choudhury, S. D., Pal, H and Mohanty, J. (2011). Supramolecular Assemblies of Thioflavin T with Cucurbiturils: Prospects of Cooperative and Competitive Metal Ion Binding. *Isr. J. Chem.*, 51, 634–645.

496. Lü, J., Lin, J.-X., Cao, M.-N. and Cao, R. (2013). Cucurbituril: A Promising Organic Building Block for the Design of Coordination Compounds and Beyond. *Coord. Chem. Rev.*, 257, pp. 1334–1356.

497. Mandadapu, V., Day, A. I. and Ghanem, A. (2014). Cucurbiturils: Chiral Applications. *Chirality*, 26, pp. 712–723.

498. Gürbüz, S., Idris, M. and Tuncel, D. (2015). Cucurbituril-Based Supramolecular Engineered Nanostructured Materials. *Org. Biomol. Chem.*, 13, pp. 330–347.

499. Shetty, D., Khedkar, J. K., Park, K. M. and Kim, K. (2015). Can We Beat the Biotin–Avidin Pair?: Cucurbit[7]uril-Based Ultrahigh Affinity Host–Guest Complexes and Their Applications. *Chem. Soc. Rev.*, 44, pp. 8747–8761.

500. Liu, W., Samanta, S. K., Smith, B. D. and Isaacs, L. (2017). Synthetic Mimics of Biotin/(Strept)avidin. *Chem. Soc. Rev.*, 46, pp. 2391–2403.

501. Hou, C., Huang, Z., Fang, Y. and Liu, J. (2017). Construction of Protein Assemblies by Host–Guest Interactions with Cucurbiturils. *Org. Biomol. Chem.*, 15, pp. 4272–4281.

502. Kaifer, A. E. (2017). Electrochemical Properties of Cucurbit[7]uril Complexes of Ferrocenyl Derivatives. *Inorganica Chim. Acta*, 468, pp. 77–81.

503. Murray, J., Kim, K., Ogoshi, T., Yao, W. and Gibb, B. C. (2017). The Aqueous Supramolecular Chemistry of Cucurbit[*n*]urils, Pillar[*n*]arenes and Deep-Cavity Cavitands. *Chem. Soc. Rev.*, 46, pp. 2479–2496.

Index